RIVER ANGELS

Shane,
Thanks for being part of the journey!

RIVER ANGELS

ROD WELLINGTON

RIVER ANGELS
Copyright © 2016 by Rod Wellington.
All rights reserved.

Cover design and interior design by VMC Art & Design, LLC.
Cover design assistance by Abstract Marketing Inc.
All photographs by Rod Wellington.

Published by Crow Books.

The uploading, scanning, and distribution of the book by any form or any means—including, but not limited to electronic, mechanical, photocopying, recording, or otherwise—without the permission of the copyright holder is illegal and punishable by law. Please purchase only authorized editions of this work, and do not participate in or encourage electronic piracy of copyrighted materials. Your support of the author's rights is appreciated.

ISBN: 978-0-9940829-3-0

This book is dedicated to River Angels everywhere.

We were able to go the extra mile,
because you went the extra mile.

Nearly every family in the Western world has a trove of Christmas traditions. Some hang a decorative wreath on the front door. Some write letters to Santa. Some faithfully attend midnight mass on Christmas Eve. Some enjoy a Christmas Day feast—turkey, gravy, stuffing, Christmas cake. Some sing carols. Most give gifts. But nearly all, in one form or another, set up a Christmas tree in a room in their home. The size of the home may vary. The size of the room may vary. The size of the family may vary. The size of the tree may vary. Some families may grow their own tree specifically for use at Christmas. Some families may purchase their tree from a scented pile of harvested pines, individually bound in chicken wire and resting domino-like in an urban parking lot—a soulless patch of pavement that remains vacant for 50 weeks of the year. Others may choose a reusable, "assembly required" department store tree-in-a-box with limbs of twisted wire and green plastic "needles" poking out from a central wooden pole. Whether man-made or natural, the chosen tree will be handsomely decorated (and personalized) with shiny strands of tinsel, dangling trinkets, and long strings of colourful lights. Quite regularly, the tree will be topped with a seasonal jewel—a smiling, pert-winged female angel; an icon of hope, a divine unification that succinctly bridges the mystery (and innocent fragility) of the spiritual world with the materialistic mortality of our own hardened plane of existence.

In the weeks prior to this holiest of birthdays, families in living rooms across the Western world turn their gaze ceiling-ward and see the welcoming eyes of a nameless angel shining down on them, wordlessly blessing their home, their thoughts, their loved ones, and affably announcing—as it is believed by many—the birth of Christ in Bethlehem more than 2000 years ago. This is Christmas tradition. This is faith.

In my youth, the same plastic tree that occupied the same corner of my parents' living room the latter half of every December was adorned with the same tinsel, trinkets, and bulbs seen on similar plastic trees in similar living rooms in similar homes all across North America. My childhood memories of the family tree are as opaque as the frost-covered storm windows that dotted the exterior of my parents' humble home in the small town of Chatham, Ontario. There remain, however, a few lucid recollections vivid enough to melt those rime-whitened panes.

Round, apple-sized bulbs made of paper thin glass were strangely punctuated with deep conical recesses. The surface of these chrome-coloured recesses was peppered with tiny indents, producing the illusion of hammered metal. Light of any colour danced wildly inside these little caves; worlds inside worlds of mirrored reflections and prismatic secrets—delectable eye candy for a youth of tiny stature.

A dozen of these bulbs had been purchased by my parents many years before my birth. They came housed in a box of thin cardboard divided into 12 little cubicles, each nesting a bulb delicately shrouded in yellowed tissue paper. They emerged from the box in glossy greens, ocean blues, cherry reds, rose petal pinks, and stainless steel silvers. Each one dangled from a wire hook that protruded from an ornate, yet slightly corroded, metal crown. When hung on the tree, they appeared as precious jewels among the green plastic needles and thin strands of tinsel. They were gems belonging to the angel on high—colourful sequins that tastefully decorated her lavishly flowing, tree-length dress.

Perhaps the most heinous of Christmas crimes in my childhood home was the accidental breakage of one of these treasured bulbs. Such a crime would not go unpunished, and punishment would not be meted by the parents, but by the child. In my case, it was self-punishment—a debilitating mix of fear, guilt, and sadness; a wash of betrayal, a loss of pride, a shattering of tradition. The fragmented remains of a broken bulb could not be passed to future generations. These remains would be discarded, trashed, forgotten—one less heirloom in a consumerist world that could use more sharing.

Never was a bulb broken while trimming the tree. That would be too convenient. Instead, they were broken when the tree was erect in mid-season splendour, in reverent moments with the eyes of family affixed upon the shimmering icon in the living room corner.

A young arm reaches up to wisp away a strand of tinsel that threatens to melt against the hot surface of a coloured light. The child's hand barely brushes a glossy green bulb, just enough to dislodge the dainty hook from the wiry branch and green needles. The bulb tumbles down over tinsel and trinkets, plunging fatefully toward a stacked pile of wrapped gifts at the base of the tree. And then, the tinkling sound of thin glass exploding momentarily fills the room, followed closely by a heavy silence of disbelief. Guilt arrives an instant later. There is a holding of breath, a seizure of muscles, a wish to be elsewhere. The child looks ceiling-ward and locks eyes with the plastic angel. She is gloriously backlit in white, her joyous expression caught in molded form, forever happy, forever overseeing—a beacon of protection and guidance. But on this cold winter night, beneath the stern, silent stares of parents and the oppressive weight of self-guilt as the child carefully picks the slivered remains of the broken bulb from the shag carpet, he realizes there is a divine flaw in the treetop cherub. She may be revered and admired from below, imbued with something holier than a molded smile "Made in China," but on this night, in his eyes, she is no saviour. She has let him down, and his view of her has been soured. She had been placed in her perch by longer arms attached to longer bodies. *They* could touch the angel. *He* could only touch the tinsel, trinkets, and bulbs. The angel, it seemed, was truly off limits. The angel, it seemed, was truly out of reach.

I'm not a religious man. I never have been a religious man and I have no intention of becoming a religious man. I never went to church as a child. Nor did my parents attend church when I was a child. Religion was never discussed in my childhood home. There was no religious imagery anywhere in the house. No crosses on the wall. No smiling Buddha on the mantel. No Koran on the coffee table. But my parents *did* own a bible; a gaudy, hardbound specimen— long, wide, heavy, and thick. Its plain white cover was embossed with regal gold text and its page edges were similarly gilded. It was a bible behemoth—oversized and underused. It held permanent residency with a hodgepodge of dress socks in the bottom drawer of my father's dresser. It was rarely seen, rarely referred to, and rarely (if ever) read. My father posits that, had it not been included as a "free bonus gift" with the purchase of a complete set of Britannia encyclopedias peddled by a door-to-door salesman in the early 1960s, my childhood home would've likely remained a bible-free zone. Curiously, my mother took special interest in the book, using it as a safe place to store funeral-related keepsakes (pressed flowers, newspaper obituary clippings, funeral programs). In the years before her death, she often browsed through these keepsakes in private, slowly reading and caressing each item of remembrance. In a way that only a book of its size could do, this bible behemoth housed a trove of tactile ties to her dearly departed friends and family. In this bible, my mother found security and sanctity. In this bible, my mother found peace.

The Bible to me is a book unfamiliar. As is the story of Jesus Christ and those who followed his lead. I cannot quote from a book I have never read, nor would I wish to. But to say that my life has not been influenced (at least in some small way) by Christianity would be a falsity. It's hard to walk through a day in our Western world without being exposed to Christian symbols and references. I'll admit I'm a complete sucker for the architectural eye candy that makes up the grand façade of a Christian church. The sight of stained glass, chiselled stone, and beaconing steeples simply arrests me. When I venture through cities for the first time, I always take a few precious moments to ogle the local churches, especially if they are dramatically sunlit at day's end. But I don't feel drawn to church service on Sundays, Good Fridays, or Christmas Eves. Words of prayer spoken behind the walls of a church have little meaning to me. These places of worship are for others, not for me. These Western world holidays and days of rest are set aside for others, not for me. These traditions, these teachings, these holy days, these buildings, these books of gospel—they are for *believers*. They are not for me.

I don't believe that a god-like entity lives in the sky, and surely not one that looks like an old Caucasian man with white hair and a white beard. If I believed that an omnipotent being existed in some realm other than the physical earth I inhabit, I certainly wouldn't believe it to have human-esque qualities. To me, such a belief is absurd at best. I don't believe humans were created in God's image. I believe that hairless apes created God in their own image. Apparently, when some unwashed ancestor of ours began searching for truth, purpose, and the meaning of life within the realm of their own consciousness, it was likely easier for them to convey their findings (presumably, the discovery of a "spirit" comprised of inconceivable energy) as something that closely resembled a human form. In many ways, humans tend to stick with what's familiar. Life is more comfortable when we are surrounded by familiar things. If our unwashed ancestor claims to have spoken (or interacted) with an entity that resembled, in their mind, a human being, it is not in my place to discredit their moment of "spiritual revelation" or "divine intervention."

I cannot prove that a man named Jesus Christ walked the same earth as me. Nor does it really matter to me that he did. What matters to me are people's beliefs—their faith, their comfort in the fact that perhaps Christ *did* walk the earth and *did* inspire people with his "enlightened" experiences (as they were told to others, and then documented in the Bible). If others take comfort in the words of Christ, that's great by me. If Christ's words give hope, truth, and purpose to those struggling in a world many view as corrupt, well, I'm fine with that. We all search for questions, and we all find answers. And sometimes during our search, we receive assistance from entities beyond our own physical body. It's always comforting to note that this *assistance*, in whatever form it comes to us, is never far away. We tend to stick with what's familiar. We are beings of attraction, and we are attracted to those things that give us comfort and peace.

Angel imagery is amazingly prolific in Western society. Likenesses of angels and angel wings are everywhere. We see them in books, magazines, movies, television, and advertising of all kinds. They are chiselled into marble headstones, sewn on the backs of black leather biker jackets, tattooed on the taut skin of mixed-martial-arts fighters, and emblazoned in gothic greyscale from collar to waist on oversized t-shirts. The image of the fallen angel resonates strongly with our inner rebel; that voice of angst and disappointment, sadly punctuated with a desperate hope for a pain-free hereafter.

There is a direct correlation between angels, death, afterlife, and rebirth. Life and death (and the journey to each) is defined wholly by *transition*—the path upon which enlightenment occurs. Our lives (and deaths) are in constant states of renewal. We are perpetually being offered second chances, additional opportunities to evolve beyond our present selves.

Along this ever-changing journey to enlightenment, we come into contact with others who recognize themselves in us. We are familiar to them in ways perhaps neither party fully understands. But what *is* understood is the attraction, the affirmation of familiarity. *That* is undeniable.

I don't believe in angels, not insomuch as they relate to Christian faith anyways. I can't believe that something watches over me while it exists in a different spiritual plane. I have no guardian angel. No angel has been "assigned" to me by a "god." There are no "messengers" that speak to me, or interact with me, or guide me. That doesn't happen to me. It has never happened to me. It will never happen to me. Angels don't point me in the right direction, or hover over my shoulder, or sit on the edge of my bed. Dead people don't talk to me. Dead people don't interact with me. Their "souls" don't communicate with me. I don't see ghosts. I don't hear ghosts. Simply put, I don't interact with entities from realities separate from the one I currently occupy.

But do I have faith in human kindness? Yes. Do I believe that humans are wholly pure and loving? Yes. Are there angels among us? I believe there are. Are they divine creatures? No. If not divine, what are they? They are of flesh and bone, as am I. They are earthly. They are of sunlight and oxygen. They are of plant and animal. They are of wind, and of river. They are *us*.

SUPAS – Day 3 – April 9, 2012 – Easter Monday

I looked around for Dale. He was hurriedly paddling his canoe downstream, already entering the lush green swamp on the same slowly flowing channel that had brought us to this remote spot on the Wolf River in northern Mississippi.

Dale's double-bladed kayak paddle slashed through the river's murky surface, flinging whitened spray in all directions. He was the only person I'd ever met who paddled a canoe with a kayak paddle. He called it his "hybrid style." He even patented the name! "It gitsya thare fastur," he said with his signature southern twang. It was quickly becoming apparent that Dale had a unique way of approaching everything, especially those things that required expediency.

Seconds later, Dale and his thick grey beard disappeared behind a stand of towering cypress trees. I shook my head and breathed deep, silently swallowing my anger…again. Day 3 was shaping up to be the most demanding of the trip, and we still had a *long* way to go.

I turned to look at Jonathan. He was dragging his 12-foot stand-up paddleboard across a narrow muddy beach to the water's edge. His soiled t-shirt and faded flowery board shorts made him look like he'd been on the river for months. In truth, his shabby appearance had taken less than 48 hours to acquire. His feet were encased in a pair of rubbery, five-toed Vibram shoes that squished through the mud and squeaked as he mounted his plastic board. He planted the black blade of his long paddle in the weedy shallows and shoved himself and the board away from the beach. In the hazy light of the late morning, I could see that a dark shadow of stubble had spread across his shaved head and face. He looked tired. We were all tired—all except Dale. Dale doesn't get tired.

"God, grant us safe passage on this beautiful day," said Jonathan, bowing his head as he rested his hands on the grip of his long paddle.

"Watch over us and guide us and give each of us the strength to forgive our own shortcomings. Amen."

Jonathan raised his head and turned to look at me. An awkward silence hung unsettled in the humid air for several moments—a little too long for my liking. Jonathan knew what I was thinking. He knew I was angry. There was no point in hiding it.

He tore his gaze from mine and turned to Robinson, who, at age 18, was the youngest member of our paddling entourage.

"Ready?" asked Jonathan comfortably, as if talking to an old friend.

"*Always!*" said Robinson without hesitation. His answer oozed an almost unpalatable positivity.

The thin blades of their long paddles cut through the murky surface of the river as they propelled their boards across the lazy eddy line and into the flowing channel.

I looked down at my muddied paddleboard and shook my head.

"How do I get myself into these *fucking* predicaments?" I blurted.

"There's an easy answer to that," said a perky and somewhat sarcastic voice in my head. "Three months ago, you said, '*Yes.*'"

The pesky voice was correct. This mess *had* started with a "Yes."

January 26, 2012

For the better part of 20 years, Vancouver, British Columbia, Canada had been my home. During that time I had a multitude of jobs—everything from hand-bombing tubs of ruined ice cream from the backs of incapacitated 18-wheelers to working in record stores selling CDs to a constant stream of mohawked punks. The one title I held the longest (and hated most) was that of Landscape Maintenance Person.

For eight long years, I slaved through a rash of soggy springs, bone-dry summers, and rain-drenched autumns. I dug holes, planted trees, laid sod, raked leaves, pulled weeds, spread soil, trimmed bushes, and cut a lot of fucking grass. I wouldn't say I *completely* hated the job (the gruelling labour definitely sucked, but it kept me in good physical shape), but when I finally quit (for the *second* time) on November

30, 2011, I vowed never to walk behind another fumes-belching lawnmower or dig another hole with a fucking shovel. (You'd think that with all the modern advances in technology, we could eliminate an activity as crudely asinine as digging a hole with a pointed piece of flattened steel attached to a fucking stick. Sheesh.) As much as I disliked that job (I think I've driven that point home by now), it did have a few advantages. One of them was a nine-month work season. Every December (based on my boss' adamant aversion to rain, and his love of spending Christmastime in tropical climes), my workmates and I would be laid off for three months. During this time, we'd collect about 60% of our monthly income through a government social safety net program called Employment Insurance, or "EI" as it's commonly called in Canada. EI ensured that the rent payments and monthly bills were adequately covered and that I could happily sit on my ass for three months and write books like the one you're currently holding. On January 26, 2012, I was doing just that when two emails from the Graybeard Adventurer himself, Dale Sanders, arrived in my inbox. I opened the second one first.

(N.B. Email and Facebook correspondence has been left unedited. Spelling, spacing, and grammatical errors have been left intact to preserve authenticity. It's worth noting that emails from Dale Sanders are often *deciphered,* not read.)

Just sent you a blind copy of an Email about a Wolf River Paddler, Jonathan Brown is organizing. We will start at Bakers pond, the source of the Wolf River, and paddle to the Mississippi. Dates: April 9 through 13 (or 14). Man, it would be great if you could join us somehow. Would you like for me to send you couple background Emails with detail? The promotion - fighting the human sex slave trade business" Rachel is working hard with us and her organization is handling marketing etc.

Very Respectfully, F. Dale Sanders

I'll admit that for several moments my mind got hung up on the words "fighting the human sex slave trade business."

"What the *fuck* is that all about?" I wondered aloud. "And what the *fuck* does 'fighting the human sex slave trade business' have to do with paddling?"

Dale's accompanying email made things a little clearer. The email had been CCed to a small group of people. Some of them were unfamiliar to me, but two of the names stood out: Dave Cornthwaite and Jonathan Brown.

On January 25, 2012 at 11:44PM, Jonathan Brown wrote:

Hey crew!!!

Just wanted to touch base.

Not sure if you heard but it looks like Dave may not be able to paddle with us due to high ticket prices plus I'm personally not hearing back from anyone that could help. If you've heard otherwise please let me know. We all know how much cooler our trip would be with a bit of Cornthwaite flava :)

Keith Cole connected me with a guy named John Hyde. He has paddled from Blackjack on to the Mississippi and is interested in jumping on board. I'll try and get in touch with him and see what's up. If you know him and have anything to say about how cool he is please let me know. Also.. ADAM… I got your friends FB friend request and message. So funny he asked us to see if Luke wanted to cruise. LUKE… I guess Adam's friend works at Shelby Farms and fully recommended we get you to paddle with us! So… are you down? JK!!!

I'll be going to a meeting at WRC (Wolf River Conservancy) on Feb. 8th and hopefully will be able to talk a bit about the trip. It's not the reason we're meeting but it may be relevant to our conversation.

I've almost got some OBS (Operation Broken Silence) details concerning our trip together and will be sending it out to you so your fully in the loop.

Plus, after that, Robinson with OBS may be contacting a few of you to connect and ask questions. He's super cool. DALE…. I know he'll want to talk to you about river info and logistics. HARRY… he'll want to talk to you about the same, in addition to rental and shuttle stuff. ADAM… he may want to talk with you to get a feel on Outdoors stance with things. LUKE…. He will probably talk with you about rental stuff.

Anyway, that's all for now. Hope you're having a great week!!!

JB

Attached for an added perspective on the proceedings was Dale's response to Jonathan's email:

Great progress - thanks so much "Jonah"; I have been calling this expedition the "Jonathan Brown Project". Looking forward to talking with Robinson, the paddle and camradery with you guys. Too bad we can't raise just $800.00 for Dave's ticket. I will make a personal contribution of $100.00, if we could come up with seven others that would do likewise. Paddling without Dave will be anticlimactic. We will just have to pull it off without him.

Very Respectfully, F. Dale Sanders

"Well, how 'bout that?" I thought aloud. "Those crazy bastards are planning a full descent of the Wolf River and they want me to join them."

Knowing that Jonathan and Dave's preferred method of river travel was by stand-up paddleboard (and figuring in the novel fact that I had never stood on a stand-up paddleboard, let alone paddled one), it took all of ten seconds of inner debate to decide that I shouldn't let an opportunity like this slide by.

"Besides," I said with a shoulder shrug, "how hard would it be to paddle a slab of plastic down a 100-mile-long river? It sounds like a good bit of fun, actually!"

There was, of course, another key motivating factor: Dave Cornthwaite was involved.

Good ol' "Corn" had an insistent knack of drumming up media interest in any adventure he embarked upon and I knew aligning myself with him would help garner more attention to my own kayaking expedition project which, if I could figure out a way to finance the damned thing, was scheduled to begin in June 2012 (five months hence). And even though I was neck-deep in a book manuscript (as well as readying myself for a semi-permanent move to Ontario), I knew instinctively that in three months' time, I'd be somewhere on the Wolf River with a bunch of smile-happy southerners and (hopefully) one ginger-haired Brit.

I pounded out a reply to Dale's email and unhesitatingly pushed the SEND button.

Dale,

Great to hear from you!

Regarding the Wolf River trip: Count me in, mate! I'll be in Ontario from March 1 until mid-June, so it won't take much to get down to Memphis from there. Looking forward to seeing everyone again! Not sure what Dave C's plans are following his March sailing trip to Hawaii. My guess is that he would be returning to England. It'd be great if he could stick around the U.S. for a few more weeks and join the paddle. I'd be willing to contribute $50 toward his ticket. I'll send him an email tomorrow and see what his position is.

Is this strictly a stand-up paddleboard trip or are kayaks and canoes welcome too? Feel free to send me that background info on the paddle.

Cheers,

Rod

Dale replied quickly and CCed his email to Dave Cornthwaite.

RIVER ANGELS

Rod, this is response to your Email earlier today about SUP'ing the Wolf. You cannot believe how happy I am to hear you can join us on the Jonathan Brown Project. As you can see, we have already been in contact with Dave. Still working it, trying to rase the $800.00. Right now we have been keeping the invitation list low key, don't wan't an unmanageable number paddlers. The original group: (1) Johanthan Brown - SUP - Apple Computer, Missionary and SUP expert (2) Harry Babb - SUP - He's the one from Ghost River Rentals. (3) Adam Hurst - SUP - He's the one from Outdoor Inc. (4) Luke Short - SUP, but looks like his College work will keep him from paddling with us. He has the SUP rental operation at Patriot Lake (5) Richard Sojourner - Canoe - He's the one that shuttled you guys from Tunica, MS. (6) Me and I will be paddling my Old Town Pac, Canoe. With you and Dave there will be seven paddlers (not counting Luke). Dave is still a possibility - Right Dave? - with you, "Jonah" and I nudging him little. Will save couple spots, in case you or Dave have someone you wish to paddle with us. Will try not going over 10, for after all, we will only be as efficient as our weakest link. After all the Wolf River is 105 miles long, through some of the most isolated swamps in the mid south. Have asked "Jonah" to put together some Eamil history, from past 30 days or so, forward them on to you, to bring you up to speed - Please include causes for "Stand Up Against Slavery" and Rachel's project support etc. (I am assuming Dave already has the history Emails). Hope I didn't miss something important, but feel, in my heart, the urgency of getting this Email released ASAP.

And just like that, I was on my way to Memphis for the second time in nine months. But before I continue with that story, let's back up a bit here and I'll talk about my *first* visit to Memphis.

August 2011

For the better part of two summer months, I had been following along online as British adventurer Dave Cornthwaite (he of slim build, ginger hair, pasty skin, unwavering determination, and endless charm—the bloke's an instant hit with women of all ages) was making his way down the entire length of the Mississippi River on a stand-up paddleboard. He'd started at the river's source (Lake Itasca in northern Minnesota), and was intent on paddling all the way to the Gulf of Mexico (100 miles south of New Orleans) in 90 days or less.

Now, for those not in the know, a stand-up paddleboard (SUP) looks a lot like a surfboard. In fact, both SUPs and surfboards share the same birthplace: the south Pacific. The premise is quite simple: you stand on the board and use a long-handled paddle (similar to a canoe paddle) to propel the craft.

Interest in stand-up paddleboarding has hugely blossomed in the past five years. The price of boards has dropped, making them affordable, and their availability has increased as well. SUPing is largely recognized as the fastest growing recreational sport in the world.

With interest comes innovation. In recent years, SUPs have undergone a gamut of changes and modifications. There are SUPs for surfing waves, speed racing, yoga workouts, lake paddling, and my personal favourite—long-distance trekking. The touring models are designed with a pointed bow to cut through the water more effectively. Additional length and thickness help support the weight of extra gear. Most long-distance boards incorporate bungee-corded areas on the front and rear decks to accommodate gear. These boards are at home on lakes, rivers, and coastal shorelines. They'll even tackle up to Class 2 whitewater!

Dave Cornthwaite embraced the stand-up paddleboard early on. In 2009, he regularly paddled one on the Kennet and Avon Canal in England whilst living aboard a brightly painted "narrowboat." He also used one to cross Lake Geneva in 2010 with Australian friend, Sebastian Terry. But before Dave reached the decision to descend the Mississippi River on a SUP, it should be noted that he'd been voraciously pursuing adventure for many years. In fact, he was making a career of it!

In 2005, our ginger-haired Brit was working as a graphic designer. He had all the lifestyle trappings of a typical young professional: good-paying job, girlfriend, mortgage, cat. One very important thing, however, was missing from Dave's life: *happiness.*

One thing that brought Dave joy was snowboarding. He loved

the exhilaration he felt when carving down a powdery slope. So one spring day, with no snow in sight, he bought a longboard (a long skateboard) and skated everywhere. Racing down hills and flying through city streets atop four tiny, whirring wheels sent a much-needed shock of self-propelled ecstasy through his being. The experience was a revelation. He loved it so much that he quit his graphic design job and decided to do something that no one had ever done: skateboard the length of the UK, from the top of Scotland to the bottom of Britain, 896 miles in total. Along the way, he planned to raise money for three charities: Link Community Development (developing sustainable teaching practices in Africa), Sailability (introducing the mentally and physically handicapped to sailing), and Lowe Trust (finding a cure for Lowe Syndrome). Despite the endless rain and disgusting blisters, the trek (and fundraiser) went so well that Dave decided to up the ante. He set his sights on Australia.

In early 2006, the intrepid Brit assembled a support crew of four (including a filmmaker) and diligently embarked on a quest to become the first person to skateboard across the Australian continent. Fourteen pairs of shoes and 3600 miles later, he did just that.

The UK and Australian treks raised a total of £50,000 for three charities and also netted Dave a book deal. His new career as an adventurer was quickly solidifying, but it would take a series of river journeys to properly cement the transition from his old life as a graphic designer.

In 2009, Dave returned to the land down under to kayak the continent's longest river, the Murray, from source to sea—a distance of 1550 miles. It was during the planning of this adventure that I connected with Corn for the first time.

I, too, was planning a kayaking descent of the Murray, set to commence about the same time that Dave's ended. My journey would be a "summit to source to sea" trek, incorporating an ascent of Mount Kosciuszko (Australia's highest peak, located about 30 miles from the Murray's source).

Dave and I had been receiving logistical river information from Rowan Privett, an Australian bloke who paddled the length of the

Murray in 2005. Seeing that our questions ran a similar gamut, Rowan proposed, through befitting email introductions ("Dave meet Rod. Rod meet Dave."), that Dave and I share with one another the information we had accrued. And thus, a fine friendship was formed.

There was a healthy dose of mutual respect between Dave and me. Both of us had crossed the Australian continent under our own power—he on a skateboard, me on a bicycle (in 2003-2004)—and by early 2010, I, too, would lay claim to paddling the longest river in Australia from source to sea.

With three long-distance adventures under his belt, and his former life as a graphic designer happily receding from view, Dave set his sights on a new horizon—an ongoing project called Expedition1000. This new endeavour would see him complete 25 separate journeys of 1000 miles or more using non-motorized means of transport. He had already crossed two off his list (skateboard and kayak) and was eager to begin another.

I finally met up with Dave face to face at the Restless bike shop in Vancouver, British Columbia in April 2011. He and his Australian buddy, Sebastian Terry (who happens to hold the Guinness World Record for breaking 24 eggs with his big toe in 30 seconds), were about to embark on a tandem bike journey from Vancouver to Las Vegas. Their goal was to do it in 14 days, arriving just in time for a speaking date at a business conference.

Our meeting at the bike shop lasted about an hour or so. Both Dave and Seb were completely down to earth; instantly friendly and generously imbued with a giddy sense of humour. Neither had ridden a tandem before and their knowledge of bicycle repair was sparse, so I, along with the accommodating staff at Restless, offered tips on how to fix a puncture and helped choose a route down the west coast. 14 days and 1400 miles later, Dave and Seb rolled into Las Vegas. Expedition #3 was in the bag.

Around the same time that Dave and Seb were pedalling their way southward, I was pushing a gas-powered lawnmower around a gated community stuffed into a well-wooded (but ultimately down-trodden) corner of a Vancouver suburb.

Cutting grass was an integral part of my landscape maintenance job, a position I'd held, off and on, for the better part of eight years. At no time during my landscaping tenure was I truly happy (in fact, I was miserable every day), but on the day in question ('twas sun-filled and warm, as I remember), a generous grin crept across my face as my mind eagerly wrapped itself around a glorious idea that had just popped into my head.

That morning, while shovelling the contents of a typical breakfast into my mouth (cold brown rice, soy beverage, spirulina, and a banana), I sat reading an interesting paddling story in Coast & Kayak magazine about a small group of kayakers who had recently floated the beautiful White Cliffs section of the Missouri River in eastern Montana. The photos alone spoke volumes. White sandstone cliffs, carved long and tall from eons of erosion, towered over the parched Montanan prairie. The sky spread wide with sapphire and puffy white as shades of tan, sand, and khaki blended into an inhospitable landscape of russet-coloured hills and spiky vegetation. I was mesmerized.

The images and words from the magazine article stayed with me throughout the workday, drifting into daydreams that happily distracted me from the inane drudgery at hand. I dredged up fond memories of past river journeys and thought long about the Murray and the Mississippi rivers, two mighty waterways I had travelled the complete length of. Those had been arduous adventures, wrought with rollercoaster emotions and unforeseen challenges. But both journeys had also immersed me deep in a natural world of beauty, wonder, and personal growth, three factors sadly absent from my work-a-day world.

But now, as of breakfast, thoughts of a new "M" river had permeated my brain. Even though I knew virtually nothing about the Missouri, other than it flowed through the badlands of Montana and emptied into the Mississippi north of St. Louis, I was quickly becoming obsessed with it. I wondered if it was the longest river in North America. Or was yet another "M" river (the McKenzie in northern Canada) the longest? River trivia tumbled into my head and

I struggled to name the longest river on each continent. The Nile, the Amazon, the Murray in Australia; those I knew. The other four? No clue. (I could see that some after-work Google research was in order.) But if in fact the Missouri River *was* the longest in North America, and if I was to paddle it from end to end, I would be able to claim that I had kayaked the longest river on two continents from source to sea.

"Hmmmm," I thought, "that would be pretty cool."

And then, without warning, the words of a thick-accented Australian bloke I'd met years prior came flooding back. When he learned I was attempting to paddle the Murray from source to sea, he asked, without hesitation, "So, mate, where do ya go from here? Are you gonna paddle the longest river on each continent?" I laughed and promptly answered with an exaggerated headshake, "Ha! Not a chance!" The idea of doing such a thing was ludicrous. "Honestly," I said to myself, "who would be foolish enough to undertake something like that?"

For years, I had been trying to come up with the idea of doing a multi-staged, self-propelled expedition, one that would allow me to take my adventuring to the next level—a level at which I could actually earn a living from adventuring, particularly through public speaking, multi-media presentations, and book sales. I was seeking a new career direction, a direction that would better reflect my passions and dreams as they related to self-propelled exploration. Missing was the precious seed that would allow this desired direction to flourish. Cornthwaite had his Expedition1000, and that was proving to be successful for him. Sebastian Terry had his 100 Things project, an ongoing bucket list that had become more of a lifestyle than a quest. He too was carving out a living.

"Maybe I should paddle the longest river on each continent," I said nonchalantly to myself as I directed the rumbling lawnmower along an arrow-straight fence line. "I've already accomplished one."

I mulled that thought over for a few short moments and then decided to up the ante.

"Why not just paddle the longest river system on each continent from source to sea?"

"Wait!" I shrieked over the roar of the mower. "That's it! That's the key I've been looking for!"

The whole idea came to me in a flash. It was one of those A-HA! moments that grabs you and will not let go; a realization that your consciousness has just lurched forward and planted you firmly in the unwritten unknown. Within a millisecond I knew there would be no turning back. It took only a millisecond more to convince the doubtful and fearful part of myself that a new direction had indeed been wrought, a defined direction that had eluded me for almost a decade, a direction defined not only by passionate adventure-seeking, but also a desire to establish a career that both reflected my self-propelled passions and my intense longing to be free from two and a half decades of unrewarding work. I knew I had worthy gifts to contribute to the world, gifts not extended through the grip of a rumbling lawnmower or the slivered shaft of a shovel, but gifts that contribute to the betterment of the world, gifts that hopefully inspire others to better themselves, to push past their imagined boundaries and cultivate the courage to take risks and to seek out happiness, purity, and truth. "*This* is my career direction," I thought to myself, smiling. "To say and do the things that best enhance my life and the lives of those around me at any given moment." By committing to such a decision, I had discovered a missing piece of my life puzzle and the discovery made me giddy.

And then, more questions arose in mind.

"Has it been done before? Has someone actually taken the initiative to descend the longest river system on each continent?"

At the time, I did not know. I had recently read (albeit briefly) about some bloke undertaking a similar project. Strangely (maybe because he was a friend of Cornthwaite's), his name came quickly to me: Mark Kalch. I resolved to find out more about Mark's project when I got home from work.

A quick Google search revealed that Mark Kalch, an intrepid explorer in his own right, was readying himself for a source to sea descent of the Missouri-Mississippi river system beginning on May 1, 2011, which meant that his expedition was already well underway.

I admit to feeling both dismayed and relieved at this discovery; dismayed because I had briefly pondered the idea of Mark and me perhaps adventuring together down the Missouri, but relieved that his intended solo descent would stay that way, and mine, whenever its fruition would come, would likely also be a solo endeavour.

Upon delving further into Mark's website, I discovered that I shared with him similar career ambitions, at least as to how they related to pursuing a decade-length, multi-stage expedition. His career as a competitive athlete, explorer, and motivational speaker seemed successful and well established.

I held back from contacting Mark straight away, chalking up the fear I felt to a lack of proper preparation and a true lapse of spontaneity. I worried that he would think I was perhaps stealing his idea of descending the great river systems of the world from source to sea. Of course, I had no such intention. If anything, I was not seeking out competition or an appropriation of his ideas. Instead, I believe I was seeking mere companionship; a planning partner or avid accomplice, one with which I could share ideas and information, and one to whom I could offer support and encouragement while he pursued his goal. I decided that contact with Mark would happen when it happened and I went about my own expedition planning and research.

Weeks later, on May 31, Mark revealed via a blog post on his website that his 2011 expedition plans had been halted. He had, in fact, never reached the Missouri River. Nor had he even left his European home. His right shoulder—injured earlier in a whitewater kayaking mishap—now desperately needed surgery, the results of which would postpone the start of his Missouri-Mississippi descent (the second of his "Seven Rivers, Seven Continents" project) to May 2012.

"Maybe there's a possibility of working together after all," I thought as I re-read his post and grimaced at the image of a bloody surgery photo. "Things *do* happen for a reason, be they unfortunate or otherwise."

And so, as ridiculous as it sounds, I began working up the courage to email Mark and inform him of my expedition plans.

I say "ridiculous" because it seems absurd to me that I should be intimidated by such a person as Mark Kalch. (I think he would comply with that thought.) Maybe I was intimidated by his ample list of expedition accomplishments and his obvious career successes. I'm not sure. What I do know is that shyness, low self-esteem, and tricks of the mind are strange things indeed. Fear can grip us tight at times when we need it not to. Surely, we are all better off without such fears. Risks, whether big or small, are always better overturned, exposed, and undertaken.

It took a full month to work up the courage to contact Mark Kalch via email and inform him of my expedition plans. Thankfully, but perhaps not surprisingly, his prompt reply was well-peppered with tones of friendliness, respect, and heaps of encouragement. A formidable bridge had been crossed and the view from the opposite bank appeared to be full of grace and good fortune.

While all of this big river research was taking place, good ol' Cornthwaite was quietly planning his next expedition—a source to sea paddling descent of the Mississippi River. Dave knew I had descended the length of the Mississippi in a canoe and a pontoon boat back in 2001 and was eager to get some firsthand information about the headwaters section in northern Minnesota. I imparted what I remembered, and asked what kind of craft he planned to use for the journey.

Ever the innovator, he replied, "A stand-up paddleboard."

At the time, no one had paddled the length of the Mississippi on a SUP. As he had done before (on a skateboard in the UK and Australia), our ginger Brit was breaking new ground. If successful, he would not only become the first person to paddle the Mississippi River 2400 miles to the Gulf of Mexico, but also smash a SUP distance record in the process. (The current record was 1808 miles.)

Back on the river research front, the task of choosing which of the seven longest river systems I would descend first was occupying my mind. Like most pursuits, the decision was based on the availability of money. I didn't have the funds to finance an overseas expedition, so it was pretty much a shoe-in that I would begin with the closest river system, the Missouri-Mississippi. I set a launch date (June 2012, a year hence) and excitedly began to grasp the enormity of the undertaking.

At a length of nearly 3800 miles, the Missouri-Mississippi river system is North America's longest. It begins high in the Centennial Mountains of southern Montana and slowly snakes its way north and east across Montana, then south through North Dakota, South Dakota, Iowa, and Nebraska, then east across Missouri, then south as it laps against the borders of Kentucky, Tennessee, Arkansas, and Mississippi before plunging wholly into Louisiana on its final run to the Gulf to Mexico. Its long, winding course is staggering to view when highlighted on a map of the U.S. Stretching nearly the whole north-to-south length of the country, and reaching west to the Rocky Mountains, this huge waterway, if measured end to end and laid east to west, would extend from New York City to Los Angeles and back to Denver.

In order to better understand the complexity of the river system, I broke it into three sections: the upper linear tributaries, the Missouri River, and the lower Mississippi River.

The upper linear tributaries are comprised of the creeks, rivers, lakes, and reservoirs that form the first 300 miles of the river system. Next is the Missouri River, which, at a length of 2321 miles, is the longest river in North America. The Missouri empties into the Mississippi River just north of St. Louis. The Gulf of Mexico is 1200 miles downstream from this confluence.

During my online research, I happened upon an 11-part journalistic series at NBCNews.com entitled *Great Escapes – Fording the Mighty Mo*. The series, written in September 2004 by American author and lifelong river man, Richard Bangs, chronicled a 10-day road trip along the length of the Missouri River. Joining Bangs was Pasquale Scaturro, who, along with fellow American Gordon Brown, made the first source to sea descent of the Blue Nile/White Nile River in early 2004. Bangs subsequently penned the bestselling book, *Mystery of the Nile* (the expedition's account), and it was through that book that I became acquainted with him and Scaturro.

In the second installment of the Missouri River series, Bangs describes their trek to Brower's Spring (the utmost source of the Missouri), high in the Centennial Mountains of southern Montana. Anthony Demetriades, a local landowner, provided key logistical assistance, turning his ranch into a home base for Bangs, Scaturro, and their media team. Demetriades, as I discovered in detail later, is a retired mechanical engineer professor at Montana State University who, for years, had been trying to persuade not only historians but also the U.S. Army Corps of Engineers that American explorers Meriwether Lewis and William Clark, in their quest to find a route to the Pacific Ocean in 1805, had not discovered the utmost source of the Missouri. Demetriades, armed with satellite maps and a wealth of research material, argued that, because Lewis and Clark had chosen to ascend Lemhi Pass instead of following the tributaries that flowed in from the east, they missed the opportunity to discover what surveyor and historian, Jacob Brower, eventually did in the mid-1890s: the true source area of the Missouri River.

The seven-mile walking route from the nearest roadway to Brower's Spring parallels Hell Roaring Creek, the first waterway on the Missouri-Mississippi river system. The creek, which passes through Demetriades' property, spends much of its time noisily racing between canyon walls before emptying into Centennial Valley. Here, it merges with Red Rock Creek and flows through the beautiful wetlands of the Red Rock Lakes National Wildlife Refuge.

Bangs and Scaturro, with help from Demetriades, decided to access Brower's Spring on foot from the north via Sawtell Peak (el. 9880 feet). A service road leads to a communication tower atop Sawtell. From there, the two-mile route along the Continental Divide ridge line, and into the cirque where the spring is located, is a much shorter alternative to the trek along Hell Roaring Creek.

Prior to 2004 (the year of Bangs' *Fording the Mighty Mo* series), few accounts of treks to Brower's Spring existed. In the mid-80s, two intrepid hikers made the journey, built a rock cairn beside the spring, scratched out their thoughts and observations on scraps of paper, and placed them in a glass jar next to the cairn.

Following a short descent on foot from the Sawtell Peak access road, Bangs and Scaturro located the storied font and its accompanying jar. Twisting off its rusty lid, they found poems, prayers, and messages left by other hikers, many of them markedly proclaiming the site as "Holy Ground." Also inside was a photocopied map from Brower's 1896 book, *The Missouri and Its Utmost Source*, and a note from John LaRandeau of the U.S. Army Corps of Engineers stating the GPS coordinates of the spring, the elevation (8809 feet), and the distance downstream to St. Louis (2619 miles).

As Scaturro deposited a message of his own and returned the jar to its place on the cairn, a rock slipped and crashed into it, shattering the jar to pieces. Saddled with guilt, but relieved that no mysterious, sacrilegious curse had instantly befallen them in this "Holy" place (cue scene from an Indiana Jones movie), Scaturro and Bangs replaced the jar with a plastic Nalgene bottle and returned on foot to their vehicle atop Sawtell Peak.

The rest of Bangs' article proved to be an entertaining read; a proper mix of American history tales and modern day parables, cleverly flourished with the author's inimitable wit. His words conjured up images of curious adventurers passing under the shadows of towering peaks en route to magical fonts that gave birth to magnificent rivers. I re-read each paragraph with deep fascination, enamoured and eager to see the Missouri's source firsthand.

Finding information online about the tributaries downstream from the river's source proved to be a daunting task. One important decision that needed nailing down was the choice of watercraft for these waterways. Satellite imagery showed Hell Roaring and Red Rock creeks to be shallow, narrow, snaky affairs—not the best place for a 16-foot sea kayak. In order to make an informed choice, I proposed a lengthy reconnaissance trip. What better way to decide what kind of watercraft to use on the upper tributaries than to actually go there and see them firsthand? Doing so would also give me a chance to view the source area in late summer which, hopefully at that time, would be snow-free. And since I was driving to Montana, I figured I might as well extend the road trip and drive over to Memphis to connect with Dave Cornthwaite as he made his way down the Mississippi on a stand-up paddleboard.

In August 2011, I loaded up my old grey Chevy pick-up, booked six weeks off work, and went in search of every river access point (bridges and public boat ramps) on the first 300 miles of the Missouri-Mississippi river system.

I started where the Missouri begins, at the confluence of two major tributaries: the Jefferson and Madison rivers near the town of Three Forks in south-central Montana. A half-mile downstream, the Gallatin River joins the flow (hence the name Three Forks). On their westward trek in 1804, Lewis and Clark scouted this area extensively. They determined that the Jefferson River was the largest of the three tributaries and pulled their boats upstream. It's worth noting that the purpose of the Corps of Discovery Expedition—led by Lewis and Clark in 1804-1805—was not to find the utmost source of the Missouri, but instead, find a water route to the Pacific Ocean.

Although the Missouri officially begins where the Jefferson and Madison rivers meet, Brower's Spring—the utmost source of the Missouri—is located 300 miles upstream. Those 300 miles are comprised of two creeks (Hell Roaring and Red Rock creeks), two lakes (Upper and Lower Red Rock lakes), two reservoirs (Lima and Clark Canyon reservoirs), and three rivers (Red Rock, Beaverhead, and Jefferson rivers). Where the Jefferson River ends, the Missouri begins. The area surrounding the confluence of the Jefferson and Madison rivers is known as Missouri Headwaters State Park. The park attracts over 25,000 visitors each year.

Montana's famous "big sky" was horizon-wide and bright azure the day I visited Missouri Headwaters State Park. An eroded, three-foot-high riverbank, covered with a well-trampled patchwork of closely cropped grass, lay 40 paces from the small parking lot. As I stood atop the bank and gazed left, the gently rippled surface of the Madison River flowed lazily past and merged unceremoniously with the equally lethargic Jefferson River. A long, V-shaped eddy line extended outward from the pointed tip of the opposite bank, indicating the placid beginning of the infant Missouri. The emergent river swept slowly past a line of exposed gravel beds, adding a brushstroke of grounding grey to the serene, sun-lit scene. Beyond the banks, a green carpet of lush, waist-high grass stretched westward for a 100 yards before yielding to a dense stand of fully-leaved cottonwoods. Beyond these trees, the dark ridgeline of distant hills painted a low, curvy contrast where the wide horizon met the expansive sky.

I listened intently and noted, with curious interest, the quiet gurgling of the mingling waters before me. From them came a tinkle, a tickle, a giggle, a lie. A snicker, a snigger, a cackle, a cry. I heard a chuckle, a laugh, a snort, and a hoot. And then, without warning, the voices were mute.

An instant later a new chorus came, louder than the first. This one was insistent, impatient, ceaseless. The flow fidgeted like an unruly youngster, uninterested in stilling itself. It gnashed loudly and uttered a gruff guffaw, "HA-HA!" Then, in a frenzied outburst, it disregarded gravity and leapt skyward—its face bubbled white with anger and oxygen. It arced and danced and spun and fell, merging seamlessly with the gently swirling mass below it. Quietly, the subdued chorus swept past the eroded banks, past the grey gravel beds, past the waist-high grass, and silently disappeared around the first of a thousand gently curved bends.

I inhaled slowly. Then exhaled. Then raised my eyes to the dark, distant ridgeline etched low upon the horizon. "Onward," I said aloud. I turned and walked the 40 paces back to The Grey Truck and worked my upstream.

A 10-minute drive brought me to the quaint town of Three Forks. The majestic Sacajawea Hotel, lit up white-bright in the mid-day sun, commanded the attention of all those who passed down the town's main street. Built in 1910, the hotel—named after the Lemhi Shoshone woman who served as interpreter and guide with the Lewis and Clark Expedition—had been recently restored for its 100th anniversary celebration. On the day I visited, the sharp-dressed staff were readying the grounds for a lavish, Western-themed wedding reception.

Just north of Three Forks, I rejoined the Jefferson at the Drouillard fishing access. The access was named after George Drouillard, a skilled hunter and language interpreter with Lewis and Clark's Voyage of Discovery. Low river levels, typical in the parched temps of mid-August, had left a generous span of bank and riverbed exposed. I walked atop the cracked mudflat and surveyed the river, imagining it swollen and swift during late-spring run-off—conditions I hoped to encounter when I returned nine months later.

Ten miles upstream, I cautiously navigated The Grey Truck across a rickety road bridge overcrowded with beer-swilling locals ignorantly oblivious to the approaching vehicle. The drunken crowd refused to yield passage and the scene turned ugly when my right palm slammed down hard on the truck's horn. As I slowly passed the angry mob, abusive derisions streamed through the truck's open windows and I silently countered their crude profanities with a profoundly rigid middle digit. "Fuck you, Bob Marley!" shouted one of the rural rednecks before flinging himself off the bridge railing and into the river below. I shook my head and rolled The Grey Truck off the bridge approach, thankful that the heated exchange had limited itself to words alone.

Beneath the bridge chaos, the Jefferson swept quietly past an equally overcrowded swimming beach. I parked for a brief moment and surveyed the surroundings. The river here was densely lined with forest. Through the only break in the thick canopy, I spied the steep rock face of a towering mountain—an excitable precursor of the majestic scenes that lay upstream and a welcome diversion from the heap of unfounded judgement lobbed at me by a herd of doughy ignoramusus.

"Onward," I said, just loud enough for the intent to be heard by my ears only.

I sped west along Highway 2, crossing open ranch land and expansive wheat fields. Near the small settlement of Sappington, the tarmac suddenly funnelled into a tight canyon. Steep limestone slopes, their white-grey faces lathered with broad stands of pine, soared skyward from the gravelled roadside. Here, road, river, and railway became a snaking trio, each marking a historical advancement in the exploration of the American West. The yawning mouth a giant cave beckoned me into the bowels of the mountains, but I fought off the tempting urge shared with early explorers and moved further upstream, past the Lewis and Clark Caverns (a popular tourist attraction) and back into open ranch land.

At the tiny town of Silver Star, I stopped to take photos of rusting railroad relics and again surveyed the placid Jefferson. Long islands, encircled with exposed gravel beds, were tranquil homes to cottonwoods and wood ducks. The river, at its present level, was perfect for worry-free kayaking. Other than a diversion dam at Parson's Bridge—put there to direct a portion of the flow into an irrigation channel—I hadn't seen any obstacles (man-made or natural) that would pose threats to paddlers, even in late spring run-off.

The beginning of the Jefferson—where the Big Hole and Beaverhead rivers merge—lay hidden beyond a parcel of private land. Far across the wide valley bottom, the Tobacco Root Mountains loomed majestically, their jagged peaks rising to over 10,000 feet.

A narrower river than the Jefferson, the Beaverhead appeared deeper and swifter as it passed through the town of Twin Bridges. Beneath the shadow of a water tower, aluminum drift boats full of anglers launched from the local boat ramp. The Beaverhead, as I came to discover, is one of America's most prized fly-fishing rivers.

I also discovered that Twin Bridges lies at the crossroads of two major east-west/north-south cross-country bicycling routes. In an effort to provide touring cyclists with comfortable accommodations (and inject some tourist dollars into the local economy), the citizens of Twin Bridges built a sizeable wooden shelter complete with hot showers, toilets, sink, an indoor eating area, and plenty of grass to pitch tents on. The best part? It's free. Users are encouraged to leave a donation which goes toward upkeep.

South of Twin Bridges, I parked The Grey Truck at a scenic roadside stop and got out to admire the view. A low backdrop of dry, undulating hills, contoured deep with coulees and dotted with sage, came to an abrupt end at Beaverhead Rock, a 200-foot-high rock formation that towered over the river valley. Resembling the head of a swimming beaver, the rock was recognized by Sacajawea during her time with the Lewis and Clark Expedition and signalled that the expedition was in the vicinity of her relatives, giving them hope that they may be able to acquire horses with which to cross the mountains to the Pacific Ocean.

At the base of Beaverhead Rock, the river snaked through a network of sloughs, ponds, and lush green marshland. I knew navigating a kayak through this morass would be challenging at best, and I jotted down a reminder in my journal to later view a satellite image of this area on Google Earth.

Arid oranges and yellows streaked the parched landscape near Dillon, a modern town with a population of 4,100 that still retains the feel of a frontier town. The first major town on the Missouri-Mississippi river system, Dillon would be my main resupply location the following spring. All food and last minute gear purchases would need to be done here before venturing into the mountains in search of Brower's Spring, so I made note of key places like grocery stores, outfitters, and the tourist information centre. A KOA campground was conveniently located on the banks of the Beaverhead, just within the city limits. Their campsites had electrical outlets and WiFi access, two things I'd be in need of when I reached this point in the expedition. The batteries for my cellular phone, laptop, and cameras (DSLR and camcorder) would all need to be charged, and Internet access would allow me to post Facebook, Twitter, and blog updates.

Just upstream from the campground, I found another diversion dam—this one located at the foot of a large, rocky cliff. The river here doglegs and tumbles over a three-foot-high diversion structure. I noted that this dam would need to be portaged. Luckily, the dam was situated next to a small park, and hauling the 16-foot sea kayak through the park to a downstream put-in would not be a major undertaking.

The landscape south of Dillon turned rugged and rusty brown as the Beaverhead carved its way past towering rock faces. A dam at Barretts Campground would necessitate another portage. My map showed that the river here branched off into a number of channels and flowed across the valley bottom and into a large network of sloughs.

Navigation through this section would be challenging. Depending on water levels the following spring, a short set of Class 2 rapids and a sizeable drop upstream of Barretts also posed potential problems.

Clark Canyon Dam and its accompanying five-mile-long reservoir marked the beginning of the Beaverhead River and the need for yet another portage. From the research I'd done prior to this recon road trip, Clark Canyon Dam would also be the spot where I'd likely switch from a packraft to a sea kayak, assuming the tributaries upstream of the reservoir were tight-turned and shallow—highly difficult settings for navigating a 16-foot boat.

I arrived at the dam and steered The Grey Truck down a steep dirt road to the head of the Beaverhead. A long, white stream of water cascaded down the dam's giant spillway and pooled briefly at the base before renewing its lengthy run to the distant sea. Two RVs were parked at a primitive riverside campground and smoke from an unattended campfire swirled skyward as fishing guides launched aluminum drift boats from a concrete boat ramp. I snapped a few photographs and noted it was here that Lewis and Clark chose to head west up Lemhi Pass, a route that would eventually take them over the Rocky Mountains. Had they gone east, they may have ascended two tributaries and discovered the true source of the Missouri River—a tiny spring emanating from a remote cirque high on a mountain slope. But that, as I stated earlier, was not their mission.

Upstream of Clark Canyon Reservoir, the braided channels of the Red Rock River snaked their way through private ranch land and public access was limited to a smattering of bridge crossings. Curiously absent here were the anglers and drift boats I'd seen on the Jefferson and Beaverhead. A short drive down a gravel side-road revealed one probable reason for their absence. Metal fence gates had been suspended from the underside of a bridge and anchored with wire into the riverbed. Whoever did that, I posited, was either trying to keep cattle off the neighbouring property, or (more likely), trying to deter paddlers and anglers from floating the river. Posted in plain sight were a half-dozen No Trespassing signs. Portaging around this obstacle would require trespassing on private property, and where there were cows, there were sure to be cowboys. And where there cowboys, there were sure to be guns. The American West was beginning to reveal an unfriendly face and it was obvious that the challenges upstream from here were only bound to increase.

I made quick stops in the small rural towns of Red Rock, Dell, and Lima as I moved south toward to the Idaho border. Services here were few and far between (two gas stations), and the river was often hidden from view, but the stunning mountainous landscape more than made up for the deficiencies.

I left the highway at the tiny community of Monida, a true frontier settlement of six homes, three outbuildings, a decrepit (but very photogenic) barn, and an inhabited train caboose decorated with Christmas lights.

East of Monida, The Grey Truck laboured up a steep gravel incline and emerged atop a level precipice. The view from this overlook was incredible. Mountain ranges on both sides hemmed in beautiful Centennial Valley. To my right, the jagged peaks of the Continental Divide loomed large and exquisitely majestic. Far below, the Red Rock River snaked along the valley bottom, entering and exiting the 12-mile-long Lima Reservoir. A dam at the reservoir's western end would necessitate a portage. Other than that obstruction, the river appeared to be flowing healthily and hindrance free.

I continued east toward the Red Rock Lakes National Wildlife Refuge and parked for the night at the turnoff to the Lower Red Rock Lake Campground. As the setting sun streaked the western sky purple and pink, I bedded down on The Grey Truck's bench seat and passed quickly into a deep sleep.

In the morning, I was greeted by Mia McPherson and Ron Dudley, two bird photographers from Salt Lake City, Utah. Mia and Ron travel to Red Rock Lakes twice yearly to photograph a huge range of migrating birds that nest and rear their young at the Refuge. When asked what brought me to Red Rock Lakes, I told them, "the source of the Missouri River," and explained my plans to kayak the river system the following year.

"I'm here to check out the upper tributaries and decide what kind of watercraft I want to use," I said. "So far, it looks like a small, inflatable packraft will work best."

"Not sure if you're aware," said Ron, pointing to a large lake to our right, "but paddling from Upper Red Rock Lake to the lower lake is prohibited at certain times of the year. You might want to check with Bill West at the refuge headquarters and see what he recommends. The headquarters are about three miles up the road at the village of Lakeview. Bill is the refuge manager and a heckuva nice guy. He'll be able to answer all your questions."

"Thanks for the tip," I said as I shook Ron's hand. "By the way, what's that round foam on your truck window?"

"That's part of a pool noodle," replied Mia with a light-hearted smile. "We rest our long telephoto lens on the foam when shooting from inside the truck. You can buy expensive window rests online, but we chose to slice open a pool noodle and slide it over the window. It's cheap and it works great!"

"Genius!" I said, smiling.

We exchanged business cards and promised to stay in touch. As they pulled away, I smiled again when I saw the letters on their truck's personalized licence plate: RAPTOR.

"How fitting," I thought.

"You're biggest problems are going to be snow pack, barbed wire fences, and ornery landowners," said Bill West as he spread a large map of the Centennial Valley across the wooden table. "This is all ranch land, and some of these landowners aren't going to be too happy about you passing through their property. If you think it might be helpful, I could make a few calls and give them a heads-up to your plans."

"Thanks," I said. "That would be great."

Bill West was a calm, personable gent in his early 60s—grey-haired, bespectacled, and dressed official in the typical U.S. Fish and Wildlife Service uniform (short-sleeved khaki shirt with crest, brown slacks, brown leather belt, black leather shoes, and an optional wristwatch). He spoke with pride and deep concern for the wellbeing of the Refuge. He'd held his managerial position here for many years, and, as illustrated by Ron Dudley's glowing recommendation, he'd earned the respect of many who regularly pass through the Refuge, and work here as well. He was precisely the guy I needed to talk to.

"Are you planning to paddle down Hell Roaring Creek?" asked Bill.

"Not sure," I said. "I've haven't seen it yet."

"I haven't heard of anyone tackling it in a boat," he said. "It definitely lives up to its name when it's running high. Gotta say, it would take some effort to paddle from Brower's Spring to where it empties into the valley bottom. Lots of debris and rock slides. Plus, it races through a pretty sizeable canyon."

He paused to look at me, and then asked, "Do you know Tony Demetriades?"

"No, but I'd like to meet him," I eagerly answered.

"This is his ranch right here," said Bill, pointing to a large parcel of land on the map. "*Roaring Creek Ranch* he calls it. It's right beside Hell Roaring Creek. I drove by there this morning. His car was in the driveway. If you like, I can give you his phone number."

"Thanks," I said. "That's mighty nice of you."

Bill looked up from the map.

"It's too bad you weren't here last week. Tony, myself, and John LaRandeau from the U.S. Army Corps of Engineers went out to Brower's Spring on horseback. John was performing a ceremony of sorts—officially recognizing Brower's Spring as the utmost source of the Missouri River. We erected a sign where Hell Roaring Creek passes under Red Rock Pass Road, right in front of Tony's ranch. You'll see it when you drive that way."

"Dang," I said, with a hint of lament. "I'm sorry I missed it. I would've loved to join you guys."

I didn't say it aloud, but I was quite excited to hear the names of people I'd read about in Richard Bangs' Missouri River account. Meeting the man who was instrumental in getting Brower's Spring recognized as the utmost source would be a highlight of this recon trip for sure!

"Do I have to cross Tony's land to get to Brower's Spring?" I asked.

"Not necessarily," said Bill. "There are three different ways to access it. Let me show you."

He leaned over the map and pointed a pencil to the spots of interest.

"The Continental Divide Trail crosses Red Rock Pass *here*, and continues up this slope behind Tony's property. Then it drops down into Hell Roaring Canyon and crosses the creek *here*. Just past this bridge, you come to a junction. Right takes you down a switchback trail that leads to the valley bottom. The Continental Divide Trail continues left and loops around to Blair Lake, then continues north to the Divide. If you take this route, you'll cross the creek *here*, *here*, and *here*, and then you'll come to another junction. Stay left and skirt Lillian Lake on its east side, then continue along the creek's east side until you reach the spring. From Lillian Lake to the spring is about four miles. There's no trail here, but you'll likely see where our horses trampled grass and weeds parallel to the creek. You might encounter hikers here as well. As long as you stick to Hell Roaring Creek, you shouldn't get lost."

He looked up from the map and smiled, "Famous last words, right?"

I laughed.

"Another way to access Hell Roaring Canyon is by following this side road," said Bill. "It leads back through the sagebrush and ends at the old Continental Divide Trail trailhead, which connects with the trail to Lillian Lake. You can park your truck here. There's a small area for tents. This is a primitive campsite on private land, so there's no pit toilet, but there *is* a small fire ring. This area is all ranch land, so be respectful. You'll likely encounter grazing cattle. Be sure to close any gates you open. If it rains, these roads can become mud pits. Use discretion before proceeding with a vehicle."

Bill pointed his pencil at a tightly stacked set of contour lines on his map—a mountain in Idaho, just east of the Divide.

"A shorter way to access Brower's Spring is via Sawtell Peak—up *here*. A public access service road leads to a communications tower atop Sawtell. You can park here and hike in to the spring. It's about two miles, but I'm guessing you'll want to hike in from Red Rock Pass Road. Correct?"

"Correct," I replied.

"Well, in that case, you're looking at about a 7.5-mile hike to the spring."

"Was there any snow up there last week?" I asked.

"No," he answered. "But keep in mind that the area around the spring has only been snow-free for about 10 days. We had a tremendous amount of snow pack this year. If you drove past Clark Canyon Reservoir, you probably noticed the picnic shelters standing in three feet of water."

"Hmmm," I said, thinking back to yesterday's drive. "Now that you mention it—I *did* see those shelters. I never thought much about it. I assumed they were boat shelters."

"Nope. Not boat shelters," he said. "That's a picnic and camping area. The reservoir is four feet above normal levels right now. That's all runoff from snowmelt, most of which came from Centennial Valley and these surrounding peaks. The Corps is releasing a fair amount of water through the spillway—you probably saw that—so levels are expected to drop over the next week. We've had record snowfall amounts this year, so, in all likelihood, you won't see a repeat of this next spring. When were you thinking about starting your trip?"

"Early June," I said.

"You might want to change that to early July. The whole Refuge could still be under snow in early June."

"Great…" I said, grimacing. "I can see how this is all going to be completely weather-dependent. When was the last time it snowed here?"

"What's today…Saturday?" he asked.

I nodded.

"Tuesday," he said. "It snowed about an inch last Tuesday."

"Great…" I said with a sigh and a shake of my head.

Bill smiled and sat back in his office chair. "You need to realize that, up here, it can snow any day of the year."

"Now," he said, pointing his pencil at the map, "are you planning to paddle through the Refuge?"

"Yes," I said.

"Well, that's where things get a little complicated. We have paddling restrictions in place throughout the spring and summer. To eliminate the possibility of disturbing nesting birds, paddlers are not allowed to pass from Upper Red Rock Lake to Lower Red Rock Lake and vice versa. There's a river marsh that connects the two lakes and navigating through it gets pretty tricky. Normally, we don't allow people to paddle from one lake to the other, but I think in your case—because you're writing a book about your expedition, and we want to be as co-operative as possible…after all, good press is better than the opposite—I think we can make an exception. When you get ready to head this way next spring, give me a call. We'll work something out."

"Sounds good," I said, shaking Bill's hand. "In the meantime, I need a place to store my kayak while I hike out to Brower's Spring. Do you mind if I store it here at the headquarters?"

"Not at all," said Bill. "I'll show you where you can put it."

I piloted The Grey Truck east along Red Rock Pass Road on the south side of the Refuge. The truck's tired suspension groaned and squeaked as it bumped over the road's potholed surface. A group of pronghorn antelope scattered as I rounded a bend near Upper Red Rock Lake. The lake, its smooth surface impeccably mirroring the vast sky above, sat easily accessible next to the road. A small campground lined a section of it southern shore. High above the lake, on the opposite side of the road, the jagged slopes of the Centennial Mountains rose high above the lake. Part of the Continental Divide, many of these peaks were over 10,000 feet.

Further on, Red Rock Creek (which empties into the upper lake), paralleled the road, disappearing between thick groves of fully leaved willows. The creek was no more than 12 feet wide and was often bisected with small islands or slowed with well-entrenched beaver dams. My map showed the creek snaking back on itself repeatedly, taking four times the distance to cover what I was able to drive by road. Lumping all of these factors together, I was able to deduce that my initial idea of using a small inflatable packraft to descend these upper tributaries was indeed the most logical choice.

The junction of Red Rock Creek and Hell Roaring Creek was hidden away on private land, but the latter was plenty accessible at the point where it rumbled through two culverts as it passed under Red

Rock Pass Road. Also here, standing tall in front of Roaring Creek Ranch, was the sign Bill West had mentioned. It read: *Hell Roaring Creek the utmost source of the Missouri River 3768 miles to the Atlantic Ocean the fourth longest river in the world.*

Far beyond the sign, the grey, rocky slopes of Nemesis Mountain, dotted green and white with pines and lingering snow patches, beautifully dominated the scenery. To its left was a sheer-sided peak rising 200 feet above the valley floor. Between this rock and the neighbouring mountain was deep-set Hell Roaring Canyon, a craggy portal through which its namesake creek noisily entered Centennial Valley.

Lack of cell phone reception meant that I was disappointingly unable to contact Tony Demetriades. The gate to his property was locked, so I resolved to try him again later, following tomorrow's hike to the source.

I located the trailhead camping spot Bill West had mentioned and spent the rest of the day organizing gear and catching up on journal entries.

Early the following morning, I cooked my breakfast of instant oatmeal, brown sugar, sliced bananas, and spirulina (a blue-green algae high in protein), and set off on foot for Brower's Spring.

The old Continental Divide Trail switchbacked up through a dense stand of pines and emerged at an unmarked junction with an unmaintained trail. This junction wasn't on my map and it momentarily created some directional confusion. I arranged some broken branches in the shape of an arrowhead pointing back the way I'd just come. Hopefully, if it was twilight when I returned, this makeshift marker might assist in finding my way back to the truck. The last thing I needed was to spend the night lost in an unfamiliar forest.

A few minutes up the trail, I came across another junction and turned east, dropping down into Hell Roaring Canyon. The creek tumbled raucously over massive boulders and mini waterfalls. The loud rush of angry water flooded out all other sounds and I knew at once I wouldn't be paddling this beast at any time of year.

Carefully crossing a railing-less bridge crudely hewn from fallen pines, I scrambled up the east side of the creek and peered out over the lush green valley bottom. Far below, Hell Roaring Creek noisily frothed its way between the canyon walls before eagerly spilling and spreading across the valley bottom in two distinct channels.

I spent an hour-and-a-half taking photos and shooting videos of the canyon, then rejoined the main trail en route to Lillian Lake.

Feeling the need to experience the creek at close range, I descended into the rocky canyon and leapt from boulder to boulder, slowly making my way upstream. Eventually, I had to abandon the idea, finding it easier to stumble through the forest on the canyon's lip. In places where the canyon walls fell away and the creek flowed freely across the rocky landscape, I chose to remove my hiking boots and wade knee-deep through the frigid stream, numbing my feet in the process.

I rested briefly on the shoreline of tiny Lillian Lake—a crystal clear alpine gem ringed with green pines. The lake's outflow joined Hell Roaring Creek as it meandered through a stand of young willows. On either side of the narrow valley, snow lingered on the tops of the highest peaks. Bright orange signs declaring *No Snowmobiles* were nailed far up the trunks of trees, evidence that people—at one time or another—had ventured into the area with motorized vehicles, regardless of the threat of avalanches.

I worked my way upstream, past meadows lush with wildflowers—the scenery ripening with pinks, purples, whites, and yellows. Two large male deer, their antlered heads lowered for grazing, stood seemingly motionless high on a steep slope.

About a mile from the source, I checked my watch—3:00pm. I'd left camp at 8:00am. Countless stops for photo ops had delayed the trek considerably. Those 90 minutes I spent in the lower canyon snapping pictures of rocks and rapids were coming back to haunt me. It had taken me seven hours to come this far, and the trek wasn't yet half finished. Once I reached the source, I'd still have a 7.5-mile hike back to the truck. Luckily, it would be downhill. But finishing before nightfall seemed almost impossible. I set a turnaround time of 4:00pm. Whether I reached the source or not by 4:00pm, I'd need to turn back, otherwise I ran the risk of having to navigate my way out of a dark forest.

RIVER ANGELS

My breathing became laboured as I trudged onward and upward. As I climbed above 8000 feet, my body began complaining loudly about both the hike and the elevation. My shoulders ached under the weight of my pack and my legs groaned as the landscape gradually became more inclined.

Bill West had told me to look for a large black rock standing out amongst the landscape, and I was relieved when I crested a small rise and saw the rock in question, shiny slick with water from a trickling stream. It was the same rock that Jacob V. Brower had seen when he ascended this valley in the summer of 1888. Brower buried a copper plate here as evidence of where he believed the source to be located. He'd disputed Lewis and Clark's 1805 claim that they had found the Missouri's source at Lemhi Pass, 100 miles to the west. 83 years later, Brower proved them wrong. Six years after that, he published his findings in *The Missouri and Its Utmost Source*.

Behind a small mound of stunted pines, I found the three-foot-high rock cairn that Richard Bangs had written about. Atop the cairn was the same sturdy army ammo box he described in his blog post. In the green metal box was a blue Nalgene bottle—the words *Greater Yellowstone Coalition* printed on its face. Was this the same bottle Pasquale Scaturro had used to replace the "sacred" glass jar he'd inadvertently shattered? I unscrewed the lid and found a cache of scribbled notes, each proclaiming a genuine love for the unique beauty of this spot. I unfolded a small scrap of paper and smiled as I read the words:

7-26-2011
Brought the Phillips family from PA up here to show off this special piece of Montana. After God made everything else He made the Centennial Valley like frosting on the cake.
Mel Montgomery
Centennial Outfitters

Also inside the bottle was a business card belonging to John R. LaRandeau, Navigation Program Support Manager, U.S. Army Corps of Engineers. I jotted down his contact information and resolved to call him when I neared Omaha, Nebraska the following summer.

I added a Zero Emissions Expeditions business card (a company name I used for my self-propelled adventures) and a scribbled note of my own to the cache, and made an entry in a well-used weatherproof journal titled *Hiker Log* located in the ammo box. The source's GPS coordinates were written in black ink on the journal's cover: 44 32.995N, 111 28449W.

As I surveyed the area surrounding Brower's Spring, I could see it was hemmed in like the steep walls of a giant sink, with all the water flowing down to the drain where I was standing. A small stream gurgled past the rock cairn and tumbled over the large black rock that Bill West had spoken of. At the base of the rock was a large pool about 12 feet in diameter and 12 inches deep. I withdrew a yellow-capped urine sample vial from my pack and filled it with source water. I'd used the same vial to carry water from the font of the Murray River to the Southern Ocean during my summit to source to sea kayaking expedition in Australia in 2009-2010. I planned to do the same with this water sample—merge it with saltwater at the Gulf of Mexico in a year's time.

After filling my water bottles and quenching my thirst, I followed the same small spring uphill for about 400 yards until it stopped abruptly at a bubbling wellspring between two watermelon-sized rocks embedded in the mountainside. It was fascinating to think that the beginning of the longest river system on the continent was actually emanating from *inside* the earth, as if the earth was giving birth to an infant river and that, over the course of my downstream journey to the sea, I would see its complete maturation. I scrambled uphill over smooth limestone and granite and emerged atop the Continental Divide.

The view was ridiculously sublime. Ebbing grey lines of mountain ranges—each ridgeline a shade lighter as distance increased—lay below me like the jagged floor of a drained ocean. I stood slack-jawed for a moment and thought, "Idaho! O had I only known you were this beautiful!" From this place of limitless splendour anything seemed possible. I inhaled deeply. Then exhaled. Then turned back to Montana and descended into the future.

For me, the immediate future consisted of returning 7.5 miles to my camp in less than four hours—easily done on flat terrain, but not so easy when you start from the spine of a continent. It was 5:00pm. By 9:00pm it would be dark. It had taken me eight hours to hike from my camp to the source. Somehow, I had to halve that time on the return.

Thunder rumbled in a distant valley as I took inventory of the items in my daypack. During my hour at Brower's Spring, the temperature had dropped considerably, so I layered up with a fleece sweater and a Gore-Tex shell. In the distance, I could see a wall of rain falling from a grey sky. I hoped my meager rations—two Clif Bars and a small bag of trail mix—would keep me from bonking. A headlamp fitted with new batteries assured I'd be able to see the immediate terrain in the dark. A GPS, along with topographic maps of the area, assured I'd know where I was, and where I shouldn't be. With water bottles full, I began the descent.

One hour. Two hours. Pink trampled wildflowers. Creek to my left. Calm in my stride. Brimming with pride. On a day spent outside. Among towering peaks. And songs from bird beaks. En route to a source. A magnificent course. That leads to the sea. And bestows upon me. Great joy and great purpose. Great goal and great surplus. Of personal strength. No matter the length. Of the mission at hand. Or the lay of the land. Or the lack of a plan. When the trail disappears. And reoccurring fears. Come on like the thoughts. Of late evening hikes. On cool summer nights. Without aid of lights. Or maps in my pocket. In a time far removed. From my present endeavour. An idea I thought clever. Then later regretted. Forever and ever. Self-trust wanes. Confidence stained. Losing my gains. Try to maintain. Some composure. Some sense. Check GPS. Consult with the maps. Fill in the gaps. Of a memory lapse. Look at the trees. Look at the sky. Release to the cosmos. A rallying cry! Choose a new route. Splash the creek with a boot. Pick up the pace. It's now a foot race! A familiar rock face. An evident place. Tis a twilight I chase. Tis a fear I erase.

Tis a night I embrace. As the day is replaced. With ominous darkness and the howling of distant wolves.

"Oh, shit!" I stammered as I stopped to check the GPS. "I didn't plan for *wolves*!"

It was now half past nine and the forest was cloaked in impenetrable blackness. The lighted screen of the GPS cast a welcomed glow on my face.

Somehow, I'd missed the main trail junction and ended up on an unmaintained side trail. I consulted the map. Of course—as is my luck in such situations—the side trail was not shown on the map. But perhaps—as though through some warped universal balance—the side trail *was* on my GPS. The main trail, however, was not. Ugh.

Directly in front of me was an unnamed creek that blindly tumbled down a steep slope to the right of the trail bed. This creek, I reckoned, would have to eventually empty out onto the valley bottom, and from there I could find my way to the main road. And from the main road, I could find my way back to the truck. Thankfully, the creek was shown on my GPS, as was a trail that ran parallel to it. The trail led to a private ranch, and the ranch had a private road that led to the main road.

I sat for a moment and weighed my options. I could attempt to hike back to the main trail and hopefully find the switchback that would lead me down the final descent to my camp. Or, I could try my luck with this creek trail. Or—and this seemed like the safest option—I could stay put and wait for daybreak.

The noise was singular at first—closer than comfortable and almost not what I thought it was, even though I knew *damn well* what it was. The noise was quickly followed by another. And another. And another. A sickening chorus of shrieks whose volume rose with my quickening heart rate.

"*Wolves*," I quietly uttered, staring wide-eyed into the black forest.

Their loud yelps and howls told me they had moved closer much quicker than I'd expected. This fact alone raised the hairs on my neck and lifted me to my feet. Staying put and waiting for daybreak was no longer an option.

In haste, I set off down the faint creek trail, immediately aware that it hadn't been maintained in a very long time. Within the first minute, my headlamp caught the ghostly outline of a huge fallen tree that lay across the trail, completely impeding any forward progression. I dove in without hesitation, gripping the thick branches with aggressive intent, pulling myself deeper into a dense wooden web. Jagged limbs scratched my face and snagged my pack straps as I became more entangled. I writhed and lunged and thrashed all about, but no matter my efforts, I couldn't get out. The chorus of wolves grew louder still. The pack, I feared, were moving in for the kill.

The panic was almost sickening, almost vomit-inducing. Trapped like a wild, angry animal—snapping, seething, snorting, scowling. Tendons tight against the skin—stretching, straining, struggling, striving to be free. Trapped in a fucking fallen tree. *This is how I die*, I thought, *detained by my own stupidity.* Eaten by wolves a mile from my truck. Surely the hecklers and hikers and haters would comment, "What a dumb fuck!"

But then, with a curse, a snap, and a leap, I fell free from the tree to the ground in a heap. I shook myself clean and continued my course, stumbling downhill with little remorse. The wolves howled on, and neared by the minute. I thought of my truck and longed to be in it. Asleep on its seat, away from the stresses of hiking to sources and endless second guesses. And then, without warning, a bridge did appear and I hurried across it and gave a small cheer. A dirt road led north through the sage and the night as the moon cracked the ridgeline, flooding the valley with light. I extinguished my headlamp as the wolves' voices ceased, and paced quickly onward, then headed due east. 'Twas eleven o'clock when I rested my head. Dreamtime was mine, like a story unread. One chapter had ended, but another I'd write. The next tale, I hoped, would not happen at night.

The next morning, I fired up The Grey Truck, followed the dirt track through the sea of sagebrush back to the main road, and ascended Red Rock Pass in search of some cell phone reception. I hoped to connect with Tony Demetriades and perhaps have a short visit before I retrieved my kayak from the Refuge headquarters. As it turned out, I found no reception at Red Rock Pass, so decided to try my luck at the small community of Macks Inn, about 20 miles east across the Idaho line.

At Macks Inn, I found a strong signal and left a voice message for Tony D, telling him of my immediate plans and interest in the Missouri's source. I also found five Continental Divide Trail thru-hikers who had started their trek in late June at the Canadian border near Glacier, Washington and were working their way south to the trail's terminus at the Mexican border in southern New Mexico. They'd already walked over 700 miles and had about 2400 miles left. I gave them a ride to a Subway restaurant down the road where we gorged on veggie subs and I satisfied an almost overwhelming craving for cold orange juice. The hikers told me they too had visited Brower's Spring, having detoured from the main trail to visit the Big Muddy's source. They'd then hiked over the Divide and descended to Macks Inn via the Sawtell Peak service road.

Unable to wait for a reply from Tony D (and slightly saddened that I hadn't made contact with him), I piloted The Grey Truck back to the Refuge headquarters, retrieved my kayak, thanked Bill West for his generous assistance, and struck eastward to Memphis, Tennessee.

It was mid-August—bone dry and hot as fuck as I passed through the raw textured beauty of the Badlands of South Dakota and stopped mid-state to view the Missouri River as it crept past the city of Chamberlain. The river here flows slowly through Lake Francis Case, a 107-mile-long reservoir flanked with spectacular cliffs and a bevy of deep coulees that drain the surrounding hills. The scenery is stark and seductive—a broad stroke of naturally delicious eye candy that doesn't overwhelm so much as it comforts and caresses with its vast silent scope of shapely terrain and its wide range of delightful surprises. I had a hunch that my next visit to Chamberlain held similar riches—precious trinkets whose enchanting elements I'd yet to unearth.

I followed the route of the Big Muddy south through Yankton, South Dakota (where the river exits the last of its reservoirs and becomes channelized), then onward through Sioux City, Iowa and

RIVER ANGELS

the Midwest metropolises of Omaha, Nebraska, and Kansas City, Missouri.

At Kansas City, I left the river and drove southwest to Wichita, Kansas. Why Wichita? Two reasons: 1) I'd never been to Kansas, and 2) a workmate and I in Vancouver had a long-running mutual gag where we poked fun at Meg White (former drummer of the White Stripes) for coming up with an asininely simple drum beat for the group's 2003 hit, *Seven Nation Army*. Our visual clue that inspired laughter at any time throughout the workday consisted of nothing more than the miming of a bent right arm striking a snare drum. One motion down, one motion up, one motion down. If you're familiar with the song, you'll know it's built around a simple bass drum beat and a simple bass guitar riff. My workmate, a devoted percussionist who ardently drools over complex rhythms and time signatures, found the song's cretin simplicity deeply unappealing, hence the mocking of Meg White's playing. Inspired (and I use that term loosely) by the song's lyrics *I'm goin' to Wichita, far from this opera forevermore*, I decided that a trip to Wichita was in order, if for nothing more than to send my workmate a text stating, "Hey! I'm in Wichita!", and then wait for the "lol" reply and some synchronized, across-the-miles, right-arm-only air-drumming of Meg White's asinine beat. And yes, that is exactly what happened. A seven nation army couldn't hold me back from telling the truth.

In a Wichita suburb, I spied a newly built (and very upscale) YMCA and navigated The Grey Truck to a parking stall. I promptly dug out a bar of soap, a washcloth, and a bath towel (yes, dreadheads own those things) and proceeded inside to procure the first shower of the road trip. Emerging cleanly shaven and wonderfully refreshed, I struck up a parking lot conversation with a burly, baldheaded gent in his mid-40s who introduced himself as Adam Knapp.

"You're a long way from home," said Adam, pointing to The Grey Truck's British Columbia licence plates. "What brings you to Wichita?"

Feeling that the White Stripes story might not resonate well with an actual Wichita resident who moments before was a complete stranger (turns out Adam is big White Stripes fan—who knew?), I decided to go with answer #1, "I've never been to Kansas, so I thought I'd check it out."

Adam Knapp seemed both impressed and slightly bewildered by my answer.

"So," he said, "you drove all the way here from Vancouver simply because you've never been to Kansas before?"

"Well, sort of…" I answered. "I'm actually on my way to Memphis to meet an adventurer friend who is currently stand-up paddleboarding down the Mississippi River from source to sea."

"Wow!" exclaimed Adam, pointing to the red plastic boat in the bed of my truck. "Are you planning to join him in your kayak?"

"You betcha," I said, smiling.

We went on to talk about Dave Cornthwaite's Expedition1000 project, especially his 3600-mile skateboard trek across Australia.

"That sounds like a massive undertaking!" said Adam. "I bet he went through a few pairs of shoes."

"14 to be exact!" I said.

Adam, as I learned, was quite a go-getter himself. Not only a devoted husband and father of two kids (11 and 14), he'd also been a veteran journalist for almost 20 years and was now editor of the Andover American, a local weekly newspaper he'd recently started. His big smile and friendly demeanour helped cement him as a well-known supporter and contributor to the Andover/Wichita arts community. We exchanged business cards, shook hands, and promised to stay in touch. (I'm happy to say we have done just that.)

In 2001, while descending the Mississippi River in a canoe and a pontoon boat with my good friend Scott McFarlane (you can read about our adventure in my first book, *Part-Time Superheroes, Full-Time Friends*), we spent a few days moored in Memphis, Tennessee and had the sincere pleasure of meeting Michele Glasnovic.

Another case where a stranger becomes an instant friend thanks to spontaneous conversation and a willingness to leave a comfort zone, I paired with Michele for the better part of three days, during which she introduced me to scores of interesting Memphians and a host of local eateries, landmarks, and urban myths. Our time together was intimate and greatly cherished—one of my favourite highlights from that river trip.

Ten years later, I contacted Michele (prior to the recon road trip) and told her I'd be in Memphis in August and that it'd be great to meet up. She enthusiastically welcomed the visit, adding that her spacious home—in Memphis' trendy Cooper-Young neighbourhood—was always open to friends.

The DeLorean in the driveway and the Mercedes on the street hinted at an increased income level well above what Michele earned at her call center job a decade prior. Her character home, painted white and vintage grey, sat on a nice-sized corner lot, lushly surrounded with a variety of plants, herbs, and flowering bushes. Hearing the low rumble of The Grey Truck approaching, Michele emerged through a backyard gate and greeted me with a heartfelt hug.

The long straight hair, the eyeglasses, the vintage clothing, the articulated intelligence, the leftist radicalism—those things were still in place. Not surprisingly, a few extra pounds had been added, balanced well with an air of calm maturity. She was still a talkative one, easily sharing details of her life—a sure sign that a level of trust still existed between us. Curiously, her lack of eye contact, something I found a tad bit annoying the first time I'd met her, was still present. Now, like then, she exuded both a gentle shyness and a brash confidence.

Many things had changed in the span of ten years, and many things had stayed the same. It wasn't my job to judge, but simply to observe and accept those changes. I hadn't been part of Michele's life these last ten years, and, quite realistically, had only been a blip of interest before then. My absence in her life (and hers in mine) had been intentional, not so much by choice, but by how our complex lives had rapidly evolved, propelling us into other life experiences and other avenues of love. A man had entered her life, and though I'd yet to meet him, I knew from her glowing description of him that he was her perfect match. The news of their partnership bruised my ego, but I'd heal. I'd get over it. Maybe.

I'd entered this visit with a distant hope that a certain chemistry still existed between Michele and me. It took only a minute to realize it did not, and a minute more to question if it ever had. My shallow hope, it seemed, would sadly need to be abandoned, forcing me instead to accept the awkward situation at hand and encourage the love she felt for Glen. Other than sulking out of Memphis with a bruised ego and a heavy heart, there were no other options. I would be a friend, and only a friend. I was certainly capable of doing just that with sincere intent.

Michele and Glen had spent most of the past year renovating their home with plans of selling it and moving west, perhaps to Portland, Oregon. The interior of the house looked amazing. Fresh paint. New fixtures. Fewer walls. More windows. They had done a lovely job. Even their two cats seemed to silently approve.

Glen, I learned, provided IT assistance to a large Memphis computer firm. His job paid well, but it had run its course and the time for change had arrived. Since our lasting meeting, Michele had returned to her hometown of Nashville where she worked as a grade school teacher. She landed a teaching job in Memphis, and, shortly after, met Glen. When they bought the house, Michele left teaching and started a home-based website design business called Spark Digital Alchemy. Her list of clients quickly lengthened and the flexibility of being a business owner allowed her to work on home renovations.

"When the house sells," she said, "we'll buy an RV, hit the open road, and work from our mobile office. We can't wait for it to happen!"

Though I tried not to show it, I was nervously anticipating Glen's arrival home from work. How much did he know about mine and Michele's time together a decade ago? Would he be jealous? Would I? Would he approve of my visit? Would he ask me to leave?

The abrupt *thud* of a car door closing disrupted the flow of pre-dinner discussion between Michele and me.

"There's Glen," she said, raising her eyebrows in mock surprise. Her humorous expression helped ease the stirring tension in my gut. My stoic charade was pointless. She knew I was nervous.

Through the back door came a balding man in his late 30s—a

little overweight, a whole lot friendly, even more intelligent than I'd imagined, and as equally nerdy as Michele (in a good way, of course). The chemistry between them was evident immediately. They were perfect for each other.

We enjoyed a fantastic vegan meal (Michele is an amazing cook), and spent the remainder of the evening talking about long-distance expeditions and DeLorean modifications. Ten years ago, curiosity and a willingness to move beyond my comfort zone had rewarded me with a cherished new friend. Tonight, that same curiosity rewarded me with another.

Glen reached out to shake my hand. I matched his firm grip and returned his smile.

"Friends are always welcome in our home," he said.

The next day, Michele and I drove downtown to check out an afternoon talk Dave Cornthwaite was giving at the Mississippi River Museum on Mud Island, a popular tourist attraction near the foot of Beale Street.

Dave happily explained how he and his stand-up paddleboarding partner, Tom Evans (a fellow Brit who joined the expedition in Minneapolis), had arrived in Memphis to a rousing reception. More than 20 members of the local Bluff City Canoe Club (in canoes and kayaks, and on stand-up paddleboards) joined Dave and Tom at the Shelby Forest boat ramp, 17 miles upstream of Memphis. Photos and videos from the gathering showed a colourful flotilla of grinning paddlers enjoying some serious summer fun on the Mighty Mississippi. Many lifelong friendships were forged that day. As a group, the club had contributed a sizeable donation to the expedition's charity, CoppaFeel!, a breast cancer awareness organization. The show of support, both on and off the water, touched Dave deeply. Memphis, it seemed, had quickly become his second home.

Michele and I chatted with Dave after his talk. It had been a crazy four months since our paths last crossed at a bicycle shop in Vancouver, and Dave, per usual, seemed to be embracing the onslaught of adventure.

Unfortunately, I never had the pleasure of meeting Tom Evans (he flew back to the UK upon arriving), but Dave's younger brother, Andy, was scheduled to join the expedition for the Memphis to New Orleans stretch. We made plans to meet for coffee the next morning.

Andy, a Royal Air Force pilot, was a charming bloke—just like his brother—and equally handsome (although I'm sure neither of them would agree on that last observation). He hadn't stood atop a stand-up paddleboard before, but was plenty keen to give it a go, even if his introduction to the sport was to take place on the biggest river in North America.

"Care to join us on the river tomorrow, mate?" asked Dave, sipping his coffee. "We're shipping off from Mud Island at 9:00am."

"Count me in!" I answered, enthusiastically.

A morning thunderstorm, complete with lightning and heavy rain, threatened to cancel our launch. Thankfully, the storm eased and moved eastward as the morning progressed, allowing a bevy of local TV reporters to properly film Dave's departure.

Always one to creatively seize the spotlight, Dave showed up in an Elvis Presley costume complete with a pompadour wig—perfect fodder for a TV news story. The outfit was linked to Dave's "Dare Dave" fundraising idea wherein he invited people to donate money and dare him to do something outlandish (within reason, of course). Members of the Bluff City Canoe Club had pooled their money and put forth the Elvis costume dare. Marrying the dare with the TV interview was a stroke of genius.

Interviewer: Can you sing like the King?

Dave: No. I try not to sing. It kills people, actually.

(laughter)

Interviewer: You look like the King. What do you have on that looks like him?

Dave: Well, I think that's almost self-explanatory, sir. I'm wearing a white jumpsuit. It's the only one we could find during Elvis Week. It's triple extra-large and way too large for me, I think—I've lost a bit of weight on this paddle. I've got a great cape, a golden necklace, sunglasses, and, of course, the wig. I don't think I'm going to take this off all the way to the Gulf.

Interviewer: Even the sideburns?

Dave: Even the sideburns.

(more laughter)

As Dave gave his interview, I introduced myself to the other paddlers joining us on the river that day. The oldest of the lot was a sprightly, grey-bearded river rat by the name of Dale Sanders. Dale had packed a lot into his 76 years. He'd won a national spearfishing title, earned a world record for holding his breath under water (six minutes and four seconds), earned a master's degree in education from Pepperdine University, travelled the world, and retired from careers in the U.S. Navy and parks and recreation. Now, as a member of the Bluff City Canoe Club and a certified river guide and volunteer with the Wolf River Conservancy, Dale showed no signs of slowing down.

Next in line chronologically was retired Memphis police officer, Richard Sojourner. Possessing an impressive grey beard of his own, Richard, too, was a certified river guide and volunteer with the Wolf River Conservancy, as well as president of the Bluff City Canoe Club and a deacon at his church. Richard was running shuttle on this day which meant he'd only be joining us for a short time on the water before doubling back to Mud Island to retrieve his truck and trailer. He'd then meet us 40 miles downriver at Tunica, Mississippi, home to a host of casinos and a river-themed museum.

Memphians Linda Weghorst and Rachel Sumner were also on hand. Linda was a member of the Bluff City Canoe Club and an avid kayaker. Rachel had helped organize the media interviews and was Dave's "assistant" for his time in Memphis, driving him around to various events and interviews.

Rounding out the group was Jonathan Brown (aka JB), a handsome, shaved-headed Apple employee in his mid-30s who had masterfully orchestrated this whole affair. A possessor of low body fat, quick wit, and a keen interest in stand-up paddleboarding, JB had heard about Dave's expedition through his connections in the SUP community. As a missionary with excellent people skills, he loved the idea of bringing folks together for a common goal—in this case, to raise awareness of breast cancer.

My new friends and I paddled away from Mud Island as the TV cameras continued to film, and, despite a light shower, everyone had permanent grins on their faces. That day, I learned the importance of a smile, especially when media was present. My mood was light—it was hard not to be with Corn dressed in an oversized Elvis costume—but even if my mood had been the opposite, it took little effort to smile when the world was watching. I'm not saying we should tailor our moods for the greater good of all, but the world seems much friendlier when it's lit with bright smiles. Like it or not, people thrive on the idea of first impressions. Media exposure is key for bringing awareness to a worthy cause. Corn and his Memphian friends understood the power of media. They also understood the power of a genuine smile, whether a camera was pointed at them or not.

"If it's not fun," said Dave, "it's not worth doing."

The weather and river were good to us that day, providing us with ample current and a swift switch from rain to sun and rampant humidity. The river itself was surprisingly warm—bathwater warm. It had been ten years since I dipped my hands in the Mississippi. It felt good to be reunited with an old friend.

Conversations on the water were personal and heartfelt. JB shared that he'd been spreading himself too thin as of late. A need for quiet time had emerged. For someone whose life revolved around the Christian concept of serving others, JB was struggling with the need for alone time, yet he knew innately that alone time was also an important means of rejuvenation. If he planned to serve to the best of his abilities, he needed to be strong. Strength, he deduced, would follow rejuvenation.

"I just hope my friends and colleagues understand," said JB as he stared downstream.

Dave and I chatted about and the ups and downs of female relationships—this from a guy who once attempted to date 100 women in 100 days with the intent of finding *the special one*. (Read his book, *Date*, to see how he fared.) We discussed how love had withered, failed, and faded from view. I told him it had been a long time since I was in a relationship—seven years to be exact.

"We stayed friends for a while afterward," I said, staring sideways at the forested riverbank, "but it's been years since we've talked."

"Do you miss her?" asked Dave.

"Yes," I answered, turning to face him. "Every single day."

Periodically, Dale would drop back from his ahead-of-the-pack river guide position and mingle with the rest of us. He quizzed me on my past expeditions and we talked long about kayaking, cycling, and backpacking. I learned he'd travelled to 56 countries, many of them with his wife of 32 years, Meriam. They'd met in the Philippines while he was stationed there with the U.S. Navy. Their three kids were adults now, moved on with families of their own.

The waterways around Memphis, especially the Wolf River, were Dale's second home. As a long-time member of the Wolf River Conservancy, his passion for environmental concerns was admirable and inspiring. He was a go-getter—determined, persistent, stubborn, unstoppable. These potent personal qualities came coupled with an overbearing nature that seemed invasive and forceful. Dale seemed to possess a need to be in charge, a need to lead the group as he saw fit. As I saw it, none of us needed to be led—this was a day paddle, not a race. I didn't welcome the competitive vibe, but then again, as one who rarely backs down from a challenge, I felt myself drawn into Dale's game. I didn't mind being told what to do, I just didn't like someone expecting me to do it. Expectations always lead to disappointments. And disappointments always lead to regrets.

I didn't talk much with Andy (Dave's brother), only touching briefly on his career as an RAF pilot. I kept my anti-military judgements to myself, electing not to say anything that I might regret later. My tongue was already sore from biting it too much. I was paddling with two adamant church-goers and two military men. I was well out of my comfort zone.

And then there was good ol' Corn, standing tall on a paddleboard dressed in an Elvis suit, silently defying criticism on numerous levels.

And then there was the river with its seemingly endless corridor of lush, unbroken foliage and expansive sandbars. Red and green buoys bobbed in the ample current as rumbling towboats pushed barges the size of city blocks upstream and down. Here, the wide Mississippi accommodated travellers of all proportions, even slow moving chatterers propelling themselves along with conversation and silent criticism.

Our destination of Tunica arrived too soon. JB and I bid adieu to Dale, Andy, and Dave and joined Richard Sojourner in his truck for the 40-minute drive back to Mud Island. Thirty-eight of those 40 minutes were spent listening to Richard talk about Joni Bishop, Ginger Leigh, and Wendy Colonna, singer songwriters from Texas and Tennessee who regularly perform house concerts at his home in Olive Branch, Mississippi.

"Y'all should think about coming to the Christmas concert," said Richard. "We do it up *real* fine—serve a home-cooked meal and ever'thing."

"Sounds like a great time," said JB.

I nodded in agreement.

The drone of Richard's voice slowly faded into the background. It merged with the rumble of the truck's engine and disappeared with the daylight as the world slid into another warm August evening here on the cusp of the South.

The day had been full. Too full. It'd been saturated with human interaction, transporting me well beyond the parameters of my comfort zone. I had paddled down a river with two military men, two church-goers, and a guy in an Elvis suit. Somehow, I had successfully navigated through a morass of silent agitation and made new friends along the way. I even found a few things to laugh about. After all, it's hard not to smile when you spend the day with a witty, ginger-haired Brit in an Elvis suit. Some people might even call that an *adventure*.

For years, I'd been trying to draw attention to my expeditions. One thing that had eluded me was media attention. I'm embarrassed to admit this, but it never occurred to me that if you want media to pay notice to what you're doing, it's always better to contact them rather than waiting for them to contact you. After all, if you're not a politician, athlete, celebrity, or convicted serial killer, there's a good chance that the media isn't monitoring your daily doings. Seriously, how the fuck would the media know if I was planning to paddle down the longest river system in North America if I didn't tip them off to the fact?

When I was kayaking down the Murray River in Australia, I arrived in the small town of Swan Hill. By sheer coincidence, I met the town's mayor at a local camera shop. He told me that a certain ginger-haired Brit by the name of Dave Cornthwaite had paddled through town months before me, and that, at a ceremony attended by local media, he had presented the key of the city to good ol' Corn. The mayor then looked me straight in the eye and said, "You should've let us known you were coming."

I must've looked confused by the comment, because his next words were: "But then, why would you?"

Exactly, I thought. *Why would I?*

Throughout that expedition, I hoped some newspaper person might happen across me and want to write a story about my journey. Sadly, that never happened. But I never went out of *my* way to see that it would happen. Corn had a knack of attracting attention to his expeditions, and *he* was no politician, athlete, celebrity, or serial killer (not that I know of, anyway). What made him so special? Why was he getting plenty of media attention? And then, I talked to the mayor of Swan Hill and it all made sense: Corn had called ahead to let the mayor know he was coming. The simplicity of the action was brilliant, and it baffles me how the idea had evaded me for this long—if you want the media to know your story, *tell them your story*. Press releases were invented for a reason.

The Grey Truck had taken a beating during the recon road trip. It had struggled to climb mountain passes in Utah and Montana, and expensive repairs had eaten into my travel budget. After Memphis, I decided to park the truck in my hometown of Chatham, Ontario and fly back to Vancouver. Doing so would leave me without a truck from mid-September until the end of November—during which time

I'd finish up the season at my landscape maintenance job—but I'd be back in Chatham for Christmas, so at least I knew there would be a vehicle there in winter.

The previous December, I came up with an idea to walk 55km around the perimeter of Chatham without actually entering the city limits. The idea began as a self-challenge, a way to top my personal best one-day walking record of 50km. With summer coming to a close, and the fact that I didn't need to be back to work in Vancouver until September 15, I decided to do the walk on Labour Day Weekend. Thanks to inspiration from ol' Corn and my new Memphis friends, the walk blossomed into a charity fundraiser. And to drum up interest in the event, I sent out a press release and did a number of interviews with local newspapers and radio stations.

The generosity displayed by the citizens of Chatham-Kent (and the fundraising efforts of the five people who joined me on the tarmac) cannot be overstated. They helped make the Go Beyond Your Limits walk a resounding success, raising $1155 for Outreach For Hunger, Chatham's food bank.

Driven by a desire to build on the success of the Go Beyond Your Limits walk, I decided to organize (with generous help from my sister, Carrie Formosa) two more fundraising events over the next eight weeks.

Autumn in Vancouver is often a soggy season, and working outside for 40 hours a week in cold, wet conditions can wear down both mind and body. It was easy to slip into a dank depression as the seasonal slog wore on. It had happened many times before.

The challenge of organizing a charity event was rewarding, far more rewarding than raking leaves and trimming hedgerows. The fundraising work gave me a goal to focus on, a sound purpose that transcended the asinine drudgery of physical labour. I was thankful for the positive distraction. I was developing other skills, skills that would propel me past the benign occupation I'd assigned myself. I was moving forward. More importantly, I was giving back.

I flew to Chatham in October to attend Jesse's Memorial Walk for Children's Safety, a tribute to Jesse Nealey, a 14-year-old Blenheim,

Ontario boy who died from injuries sustained in a bicycle accident. Jesse's mother, Annette Nealey, had participated in the Go Beyond Your Limits walk and we had struck up a friendship. The sad news of Jesse's death came six days after that event.

Thanks to local media coverage and an outpouring of generosity from the Chatham-Kent community (including Jesse's family and friends who participated in the walk), the event raised over $6800 for a bicycle safety program at the Chatham-Kent Children's Safety Village.

After eight long years, I parted ways with my landscape maintenance job at the end of November and immediately flew to Chatham to promote Go Beyond Your Limits 2—a 100km, 24-hour walk on a 400-metre oval track at a local high school. Despite the chilly December weather, over two dozen people joined the walk and the event helped raise $2025 for Outreach For Hunger.

I stayed in Chatham for the remainder of December and flew back to Vancouver on New Year's Day. My old employer was nice enough to grant me "layoff" status (at least on paper) and I was able to collect bi-weekly Employment Insurance cheques which paid the bills while I worked on the manuscript for my first book, *Part-Time Superheroes, Full-Time Friends*.

Around this time, a post from a British adventurer named Paul Everitt appeared on my Facebook News Feed. He was selling his four-wheeled Bikecar for $500—a steal at that price. The Bikecar was currently being stored in Eugene, Oregon, about six hours south of Vancouver. Judging from the accompanying photos, it looked pretty darn cool. I seriously considered purchasing it. There was one major problem, however. My truck was in Ontario. Retrieving the Bikecar from Eugene would mean renting a vehicle, and that expense would far exceed the worth of the bike. I decided to shelve the idea and ponder my options.

A few days later, Dave Cornthwaite sent me a message via Facebook. The first half of the message was addressed to me. The second half had been sent out as a Facebook post.

Rod, my man! Don't suppose you know anyone in the Pacific Northwest who might have room to store a four wheel bikecar?!!! Just been offered it but need to find storage before I can take it on the road! Dave

Help! Does anyone living in the Pacific Northwest have space for in a shed or garage to store a four-wheeled bicycle? I've been offered this for a journey (date tbc) but it needs a new home until I can take it on the road. Currently it lives in Eugene, Oregon, and it's already been on quite a journey with Paul Everitt, check out www.going-solo.co.uk.

"Well, so much for buying it," I thought, frowning. "But at least ol' Corn will get some use out of it. I'll send him a reply and offer some help."

RW: Mr. C! Have you sorted out the bike car storage yet? I see on the post comments there was a guy named Matt Ivey who offered a space. If it's not yet sorted, I could make some calls and see I can find some storage.
DC: Hey dude, nothing sorted yet. A couple of people have expressed interest but nothing solved yet. Really interesting contraption, two people can sit and pedal, quite amazing!! Any ideas you have buddy, that'd be ace. ☺

So, what's a Bikecar and how did it get to Eugene, Oregon? Glad you asked. Picture a small, four-passenger car (like a Mini) with a robust, rectangular aluminum frame. Now, take away the shiny body (but keep the front and rear seats) and replace the wheels with bicycle wheels. Install a bicycle drivetrain (complete with a super long chain) for the driver and another for the passenger. Rig it so the passengers in the rear seats can pedal as well. Change the steering wheel to a handlebar and add a complimentary steering assembly. Insert one, two, three, or four willing participants and "Voila!"—instant, eco-friendly propulsion.

A quick visit to Paul Everitt's website revealed that he'd built the Bikecar entirely by hand in his mother's garage in Wales. Videos on YouTube showed the efforts he went through to procure the needed parts, weld the frame, sew the seat covers, and assemble the thing. This was *one* driven adventurer!

RIVER ANGELS

After the obligatory test drives and tweaks, Paul and three friends pedalled the Bikecar through western Europe for 33 days. They stowed their gear in a fibreglass, rooftop cargo carrier (like you'd see on a car) mounted atop a hand-built, two-wheeled trailer which they towed behind the bike. Brilliant!

Building on the success of that adventure, Paul set his sights on crossing Canada from Halifax, Nova Scotia to Vancouver, British Columbia in the summer of 2011.

The Canadian crossing was an arduous journey, to say the least. Paul pedalled most of it solo, grinding his headwind-hindered way across the beautifully desolate Prairies and pushing the 300 lb. Bikecar up steep highway grades in the Rocky Mountains. Along the way, he was befriended by hordes of Canadians, an experience that cemented a deep love for Canada and its gracious citizens.

Upon reaching the west coast, he boarded a ferry to Vancouver Island, visited the city of Victoria, returned to the mainland, pedalled south to Washington state, and ran out of money in Oregon—prompting his decision to store the Bikecar in Eugene and return to his plumbing profession in the UK.

On January 27, 2012, I contacted Paul via Facebook message. Here's our conversation from that day:

RW: *Hey Paul. Have you sorted out the bike car storage yet? I received a PM from Dave Cornthwaite earlier today asking if I knew of anyone locally who might have storage. (I live in Vancouver, BC.) I see on Dave C's FB page storage request post comments there was a guy named Matt Ivey who offered a space in northern CA. If it's not yet sorted, I could make some calls and see if I can find some storage.*

PE: *Hey Rod, Its not sorted yet. I have till 31st of Jan. I have donated the bikecar to Dave to use on his trips. Just have to find it a home till he gets out there.*

RW: *Have you been in touch with Matt Ivey?*

PE: *No. Not ye. I will do so soon. Be good to sort it asap. Ha. Sending him a message now.*

RW: *It's in Eugene, right?*

PE: *Yep its in Eugene.*

RW: *Is there anywhere else in Eugene that it can be stored? Have you tried contacting a local bike shop there?*

PE: *Yep. I have had no replys. Eugene is a very weird place. I found when I was there that there not a helpful bunch of people.*

RW: *I've been to Eugene a few times myself. It's a unique little place, quite odd memories of it in fact.*

PE: *Very odd people. Haha. They don't open there arms to strangers that's for sure. I have a feeling the Bikecar will end up in a bad place if I don't get it sorted by the 31st.*

RW: *However, there is a major festival that happens there annually and I think someone associated with it may be able to help. I need to do a little research and see what I can come up with.*

PE: *I appreciate all the help I can get. Would be nice to see someone else take it on a new adventure.*

RW: *Have you been in contact with any bike people in Portland regarding this?*

PE: *I have been emailing people for the last month there.*

RW: *Hmmm. Very odd. We've entered a bike car void…*

PE: *Its like the bikecar has run out of loving. Haha. Its very strange.*

RW: *Let me look into this and I'll get back to you. Let me know how things go with Matt Ivey.*

PE: *Will do. I have messaged him so just waiting on a reply.*

RW: *One last thing. What happens to the bike car after the 31st? Is it in storage now? Can that be extended? Are you paying for storage there?*

PE: *The guy was meant to be storing it for free. Now hes had a mood swing and is tring to cash in on it. Hes asking for $50. If I don't pay on the 1st it becomes his for scrap he says.*

RW: *$50 per month? Per year?*

PE: *A month from what I can gather.*

RW: *Do you want to pay it?*

PE: *Right now I don't have the means to pay it and he wasn't ever going to charge me. This is new this week.*

RW: *Hmmm, new this week. Interesting. Has Dave C offered any money?*

PE: *Nope. I just want to see the bike car end up in a nice home.*

RW: *The next question is: do you even want to keep storing it there or find somewhere else to store it? Obviously, free storage would be best. The guy sounds a little shady.*
PE: *Yep very shady guy. $50 seems like alot. If it was in canada I know many people who would takke her in.*
RW: *Dude, I would take it if it was here. I know you offered it for $500 which seems like a steal. I'm really surprised no one bought it.*
PE: *Everyone who got in touch wanted it for free. They said it wasn't worth $50. That hurt. Haha.*
RW: *I don't want to see the bike car go to scrap and I don't want to see you giving money to someone you don't fully trust. I'll make a few calls and see what happens. Hopefully things with Matt in CA will work out. I'd like to see Dave use it for a future expedition. It's a really unique vehicle and it could tell a load of stories, I'm sure. We'll get this sorted out, mate. Stay in touch and let me know of any updates. I'll do the same. Cheers.*
PE: *Cheers Rod. This means a lot.* ☺

Later that day…

RW: *Paul, here's where I would start in Eugene. Not sure if Icky's Teahouse still exists (there seems to a current Eugene address for the place). I was there like 20 years ago and it used to host live bands. I was on tour with a Vancouver band called Pork Queen. It was run like a collective, punks and anarchists involved in decision making, and it seemed like a cool place. This seems like a good place to start. There has been recent activity on Icky's FB page (link below). I suggest sending them a PM regarding the bike car. I'll do the same. Here's another lead. If you haven't already, shoot off an email to the organizers at the Oregon County Fair. There's a good bet that someone from here will help you out. I have a good feeling on this.*

I turned away from the computer and stared out the window. The cold, January rain poured down, spreading a gloomy gloss across the city. Winter in Vancouver can be a depressing time. I'd been back not four weeks and already I longed for a reprieve—a *permanent* reprieve.

"Damn it!" I shouted. "I could really use The Grey Truck right now!"

The only thing stopping me from driving to Eugene to get the Bikecar myself was the fact that my truck was in Chatham, Ontario, 3000 miles away. I didn't foresee a need to have the truck in Vancouver because I wasn't planning to stay long in British Columbia anyway. With the recent success of the charity walks, Chatham was revealing far more opportunities for creative pursuits than Vancouver. It seemed a permanent return to Ontario was brewing. The signs had revealed themselves one by one. My seasonal landscaping job had ended. The manuscript for *Part-Time Superheroes, Full-Time Friends* was nearing completion. The Missouri-Mississippi kayaking expedition was slated for a June start. And, to drive the point home, my kayak was stored in Chatham. A decision had been made. It was time to move on.

There was one little problem with these well-intended travel plans: lack of money. I had no idea how I was going to fund the upcoming kayaking expedition. My Employment Insurance payments would be slightly disrupted due to the relocation, but at least a steady income flow would happen until June. Once I was in the U.S., however, I wouldn't be able to claim for payments. My file would be suspended and eventually closed. How then, would I feed myself for four or five months while on the river? I reassured myself that I would earn the money somehow. Money always seemed to appear in my life exactly when it was needed, and not a moment too soon. Money and *emails*, that is. Especially emails from grey-bearded, church-going, former naval officers from Memphis, Tennessee. They seemed to have a way of finding me and enticing me with outlandish ideas, like piloting stand-up paddleboards through snake-infested swamps in the American South. And I, for some inexplicable reason, seemed to have a strange desire to join them.

SUPAS – Day 3 – April 9, 2012 – Easter Monday

"Where the *fuck* is Dale?!" I shouted through the gloomy twilight.
I could see ol' Corn standing tall on the riverbank, all Britishly ginger-haired and cheery. He held a flashlight in one hand and his ever-present GoPro camera in the other.

"It's nice to see you too, mate," he said, smiling.

I blew off his sarcastic greeting and repeated my question.

"Seriously, where the *fuck* is Dale?"

"He's in his tent. Is everything okay, buddy?" asked Corn.

"No! It's *far* from fuckin' okay!" I shouted as I forcefully threw my paddle in the long grass and stepped onto the bank.

Corn's flashlight lit up my scowling face. I jabbed a finger back in the direction I'd just come.

"I've been out in that *fuckin'* swamp for nine *fuckin'* hours, pulling that *fuckin'* board over fallen trees, beaver dams, and a thousand *fuckin'* lily pads! And somewhere in the last hour, Dale decided to *fuck off* and find his own way to camp. None of us have seen him since. *Fuck, man! We've been worried sick about him!*"

"I just spoke to him, mate," said Corn, calmly. "He's fine. Tired, but fine."

"Good…"

The rest of my sentence never emerged. I took a deep breath and looked at Corn.

"What happened out there, buddy?" he asked.

"*Fuck me…*" I replied, shaking my head. "What *didn't* happen?"

January 26, 2012

(N.B. Email and Facebook correspondence has been left unedited. Spelling, spacing, and grammatical errors have been left intact to preserve authenticity. It's worth noting that emails from Dale Sanders are often *deciphered,* not read.)

On January 26, 2012 at 10:47AM, Dale Sanders wrote:

Just sent you a blind copy of an Email about a Wolf River Paddler, Jonathan Brown is organizing. We will start at Bakers pond, the source of the Wolf River, and paddle to the Mississippi. Dates: April 9 through 13 (or 14). Man, it would be great if you could join us somehow. Would you like for me to send you couple background Emails with detail? The promotion

- fighting the human sex slave trade business" Rachel is working hard with us and her organization is handling marketing etc.

Very Respectfully, F. Dale Sanders

I'll admit that for several moments my mind got hung up on the words "fighting the human sex slave trade business."
"What the *fuck* is that all about?" I wondered aloud. "And what the *fuck* does 'fighting the human sex slave trade business' have to do with paddling?"
Dale's accompanying email made things a little clearer. The email had been CCed to a small group of people. Some of them were unfamiliar to me, but two of the names stood out: Dave Cornthwaite and Jonathan Brown.

On January 25, 2012 at 11:44PM, Jonathan Brown wrote:

Hey crew!!!

Just wanted to touch base.

Not sure if you heard but it looks like Dave may not be able to paddle with us due to high ticket prices plus I'm personally not hearing back from anyone that could help. If you've heard otherwise please let me know. We all know how much cooler our trip would be with a bit of Cornthwaite flava :)

Keith Cole connected me with a guy named John Hyde. He has paddled from Blackjack on to the Mississippi and is interested in jumping on board. I'll try and get in touch with him and see what's up. If you know him and have anything to say about how cool he is please let me know. Also.. ADAM… I got your friends FB friend request and message. So funny he asked us to see if Luke wanted to cruise. LUKE… I guess Adam's friend works at Shelby Farms and fully recommended we get you to paddle with us! So… are you down? JK!!!

I'll be going to a meeting at WRC (Wolf River Conservancy) on Feb. 8th and hopefully will be able to talk a bit about the trip. It's not the reason we're meeting but it may be relevant to our conversation.

I've almost got some OBS (Operation Broken Silence) details concerning our trip together and will be sending it out to you so your fully in the loop. Plus, after that, Robinson with OBS may be contacting a few of you to connect and ask questions. He's super cool. DALE.... I know he'll want to talk to you about river info and logistics. HARRY... he'll want to talk to you about the same, in addition to rental and shuttle stuff. ADAM... he may want to talk with you to get a feel on Outdoors stance with things. LUKE.... He will probably talk with you about rental stuff.

Anyway, that's all for now. Hope you're having a great week!!!

JB

Attached for an added perspective on the proceedings was Dale's response to Jonathan's email:

Great progress - thanks so much "Jonah"; I have been calling this expedition the "Jonathan Brown Project". Looking forward to talking with Robinson, the paddle and camradery with you guys. Too bad we can't raise just $800.00 for Dave's ticket. I will make a personal contribution of $100.00, if we could come up with seven others that would do likewise. Paddling without Dave will be anticlimactic. We will just have to pull it off without him.

Very Respectfully, F. Dale Sanders

"Well, how 'bout that?" I thought aloud. "Those crazy bastards are planning a full descent of the Wolf River and they want me to join them."

Knowing that Jonathan and Dave's preferred method of river travel was by stand-up paddleboard (and figuring in the novel fact that

I had never stood on a stand-up paddleboard, let alone paddled one), it took all of ten seconds of inner debate to decide that I shouldn't let an opportunity like this slide by.

"Besides," I said with a shoulder shrug, "how hard would it be to paddle a slab of plastic down a 100-mile-long river? It sounds like a good bit of fun, actually!"

There was, of course, another key motivating factor: Dave Cornthwaite was involved.

Good ol' "Corn" had an insistent knack of drumming up media interest in any adventure he embarked upon and I knew aligning myself with him would help garner more attention to my own kayaking expedition project which, if I could figure out a way to finance the damned thing, was scheduled to begin in June 2012 (five months hence). And even though I was neck-deep in a book manuscript (as well as sorting out the Bikecar details and readying myself for a semi-permanent move to Ontario), I knew instinctively that in three months' time, I'd be somewhere on the Wolf River with a bunch of smile-happy southerners and (hopefully) one ginger-haired Brit.

I pounded out a reply to Dale's email and unhesitatingly pushed the SEND button.

Dale,

Great to hear from you!

Regarding the Wolf River trip: Count me in, mate! I'll be in Ontario from March 1 until mid-June, so it won't take much to get down to Memphis from there. Looking forward to seeing everyone again! Not sure what Dave C's plans are following his March sailing trip to Hawaii. My guess is that he would be returning to England. It'd be great if he could stick around the U.S. for a few more weeks and join the paddle. I'd be willing to contribute $50 toward his ticket. I'll send him an email tomorrow and see what his position is.

RIVER ANGELS

Is this strictly a stand-up paddleboard trip or are kayaks and canoes welcome too? Feel free to send me that background info on the paddle.

Cheers,

Rod

Dale replied quickly and CCed his email to Dave Cornthwaite.

Rod, this is response to your Email earlier today about SUP'ing the Wolf. You cannot believe how happy I am to hear you can join us on the Jonathan Brown Project. As you can see, we have already been in contact with Dave. Still working it, trying to rase the $800.00. Right now we have been keeping the invitation list low key, don't wan't an unmanageable number paddlers. The original group: (1) Johanthan Brown - SUP - Apple Computer, Missionary and SUP expert (2) Harry Babb - SUP - He's the one from Ghost River Rentals. (3) Adam Hurst - SUP - He's the one from Outdoor Inc. (4) Luke Short - SUP, but looks like his College work will keep him from paddling with us. He has the SUP rental operation at Patriot Lake (5) Richard Sojourner - Canoe - He's the one that shuttled you guys from Tunica, MS. (6) Me and I will be paddling my Old Town Pac, Canoe. With you and Dave there will be seven paddlers (not counting Luke). Dave is still a possibility - Right Dave? - with you, "Jonah" and I nudging him little. Will save couple spots, in case you or Dave have someone you wish to paddle with us. Will try not going over 10, for after all, we will only be as efficient as our weakest link. After all the Wolf River is 105 miles long, through some of the most isolated swamps in the mid south. Have asked "Jonah" to put together some Eamil history, from past 30 days or so, forward them on to you, to bring you up to speed - Please include causes for "Stand Up Against Slavery" and Rachel's project support etc. (I am assuming Dave already has the history Emails). Hope I didn't miss something important, but feel, in my heart, the urgency of getting this Email released ASAP.

With my foreseeable future plans now laid, I turned my attention back to the Bikecar. I drafted an information email (see below) and sent it to 30 bicycle-friendly organizations in the Pacific Northwest, everything from Vancouver's Momentum Magazine to the Black Butte Center for Railroad Culture near Weed, California.

Sent January 28, 2012 at 3:59PM:

Greetings,

My name is Rod Wellington. I live in Vancouver, BC. I am writing today to see if you can help with a small dilemma that I am attempting to rectify.

An acquaintance of mine, British adventurer Paul Everitt (www.going-solo.co.uk), is currently storing his bike car (picture in link) in Eugene, Oregon. The place he is storing it at (a person's residence) is going to start charging him $50/month for storage starting Feb 1. Paul had been storing it there for free but now the owner wants to start charging him. He's been told that if he doesn't pay, the bike car will be scrapped. That would be a pity as another British adventurer and friend, Dave Cornthwaite (www.davecornthwaite.com), is planning to use the bike car for a future expedition. Paul's plan is to continue to store it in Eugene, albeit at a different location. He is also open to storing it elsewhere along the west coast. Both Paul and Dave are currently in England and unable to deal with the bike car directly. If you know of anyone in the Eugene area, or elsewhere in the Northwest, or northern California for that matter, that has space to store the bike car and are willing to help, please let me know. (It would be most ideal and convenient to find someone willing to store it in Eugene, seeing that the bike car is already there.) Feel free to have them message me on Facebook or via email (links below). They can also message Paul Everitt via his Facebook page (link below).

Cheers,

Rod Wellington

The next day (January 29), I received a hopeful reply from David Mozer from the International Bicycle Fund.

Hi Rod,

I am forwarding this message to some folks in Eugene who are sympathetic to cycling. Being the whole to town tends to be, hopefully these folks and their networks can find a large enough corner to store the "bike car" either temporarily (to buy time) or until Cornthwaite arrives. Hopefully you will hear from someone soon.

Regards,

David Mozer

International Bicycle Fund (www.ibike.org). Promoting sustainable transport and understanding worldwide. A non-profit organization. Donations are welcome.

Meanwhile, in Memphis, Jonathan Brown and his Wolf River friends were busy hammering out SUPAS details.

On January 29, 2012 at 7:43PM, Jonathan Brown wrote:

ALOHA EVERYONE!

This is the first all-inclusive email being sent out! If there's anything I leave out please let me know.

The purpose of this email is to get everyone on the same page, update everyone, welcome new team members, introduce and connect, as well as simply get everyone stoked for what's about to go down this Spring!!!

KEITH (Kirkland), TOM (Roehm), & KEN (Kimble):
I've CC'd you guys to keep you in the loop. I hope you don't mind. I think

this will naturally benefit WRC and is in line with some of WRC's goals for 2012.

RACHEL (Sumner) & ROBINSON (Littrell):
I've got you guys CC'd into this so you can start talking to everyone you need to and from this point we'll all try to keep you included in on our communication. We also want to be kept in the loop with your side so please keep us up to date. :)

PADDLERS:
Just below are links to some documents I've put together that will need each paddlers attention. Please click on these links. I've given you permission to edit each one. This will help us communicate aside from email and meetings and help OBS (Operation Broken Silence) get some key info they will need. I will try and get another meeting together with everyone soon! Stay tuned!!!

RIVER SECTION & STATION INFO (document)

PADDLE EQUIPMENT & GEAR (document)

PADDLE TEAM INFO (document)

INTRODUCTIONS:
I'd like to introduce Rachel and Robinson! Most of you know Rachel already but none of you know Robinson. Rachel is the woman who's driving Operation Broken Silence's efforts with SUPAS (Stand Up Against Slavery) and Robinson is the hands and feet. You will probably hear a lot from Robinson as he may need to connect with each one of you for various reasons.

* If you don't know much about Operation Broken Silence (OBS) please feel free to hit up Rachel and Robinson and check out www.operationbrokensilence.org.

WELCOME NEW PADDLERS!!!

RIVER ANGELS

I'd like to officially welcome Richard Sojourner and Rod Wellington to the SUPAS paddle team!!!!

Richard Sojouner was a part of Dave Cornthwaite's time in Memphis. He's going to be representing Bluff City Canoe Club (BCCC) and in addition to being a great paddler, is pretty much one of the coolest concert promoters on earth! :)

Rod Wellington is a new and good friend. He's a man with a plan and purpose. He and Dave met a while back and he joined Dave and others on the paddle from Memphis to Tunica. Such a great guy with tons of experience. The best thing about this is that he does adventures for a living and for a purpose every time!!! Having Rod on board will spice up our trip as well as broaden our reach and further help reach our goals.

Check out what Rod's all about at:
www.zeroemissionsexpeditions.com (N.B. Website now defunct. Current website: www.rodwellington.com.)

SUPAS UPDATE:
Things have been coming together! After having a meeting with Rachel and Robinson I fully realized how much potential this event has and how much it could benefit OBS, WRC (Wolf River Conservancy)*, and so much more!*

You all will learn more about what is going on and OBS will be connecting with you to help and get answers to questions they may have. As we focus on the expedition itself, OBS will be taking care of everything else. With that said, here's some of the ideas they are working on....

- Splitting up the Wolf into 10 "theme-based" sections. Every section carrying a theme of some sort incorporating competitions and asking for donation-based involvement.
- Promoting new WRC memberships by offering "Special Edition" SUPAS/WRC Membership packages through the week of the paddle.

- Save A Slave Campaign that will run through the week. Street teams raising money based on expenses needed to rescue, rehab, and re-enter victims.
- Last day closing event with music, food, games, raffles, and paddle demos.

These are just four ideas milling around. There is so much more in the details that will get you super stoked. As plans get clearer then OBS will fill us in.

The main focus for them right now is clearly defining partnerships and roles, trumping up sponsors, building up our communications and online network, and developing the marketing and promotional strategies. If you have any ideas or could offer any help personally or through connections please don't hesitate to get in touch with these guys.

Dave Cornthwaite: You thought I forgot about you didn't you? Welp…. guys… IF Dave gets $800 then he's here with us. Any ideas… aside from robbing a bank?

OK, that's all from me. Let the ball begin to roll!!!

PS: Please don't email me at my Apple email address…. Apple tends to frown on this. :)

MAHALO!!!
Jonathan Brown

On January 29, 2012 at 10:28PM, Dale Sanders wrote:

Great effort - Jonah. We are so fortunate to have a community organizer, such as you, to run worthwhile projects. I did some editing on each three below. We still need $650.00 of the original $800.00 for Dave's ticket. Wish I was a fund raiser. (Instead of Hell Raiser)

On January 30, 2012 at 3:37PM, Robinson Littrell wrote:

Hey guys! It is great to finally be in touch with you all! I'll be your OBS contact until Rachel gets back in town.

I'm newly promoted Director of On-Campus MOVE at OBS. I work with high school and college groups who are interested in human trafficking and genocide. I'll be the one coordinating with you guys about times and dates and whatnot as we get sponsors and partners and people involved. Thanks for being a part of this! Yall are studs!

Meanwhile, back on the Bikecar front, I touched base with Paul Everitt (via Facebook on January 30) to discuss storage options.

RW: *Hey Paul, Bike car storage update. I sent out 30 emails/messages/FB posts to various bike shops, newspapers, bicycle organizations in the Eugene area. No reply yet. Hoping to hear back on them soon. Let me know if things change at your end. Will do the same here. I sent a copy of this update to Dave C as well. Cheers.*
PE: *Morning Rod, Thanks for your hard work. I emailed a few people in the early hours too. Its just a waiting game now.*
RW: *Hey Paul, do you have the contact info for the guy in Eugene that is storing your bike car? I was thinking of contacting him to let him know that I am sourcing out a solution nearby. Maybe we can buy some more time. I have a lead for possible storage in northern California, but that leaves the problem of getting the bike to Cali. Unfortunately, my truck is in Ontario, otherwise I'd drive down and get it.*
PE: *Hey Rd, My "0" butt0n seens t0 have st0pped w0rking. Haha. T0 sec's. His name is Jeff. His email is jeff@***********.com. His Office: ***-***-**** and Cell: ***-***-****.*
RW: *Is his last name Lozar, by chance?*
PE: *Yes. Do you know him? Haha.*
RW: *No, just did a little research.*
PE: *He's posted a few times on FB.*
RW: *I've had a few nibbles from a couple bike advocacy groups, saying that they are passing the request for storage around.*
PE: *Same here.*

RW: *And i got a good lead from a friend in Santa Cruz.*
PE: *I have been talking to Hannah she's something to do with a bike work shop there.*
RW: *Any leads with Hannah?*
PE: *Hot yet, I replied about this time yesterday. I wake up every hour and check emails. *Not. Sorry for bad spelling and grammar…. Eyes are still half asleep.*
RW: *I will be contacting some bike people in northern Cali today. They may have some space. Trouble is, getting the bike to Cali.*
PE: *Would only take a monthto cycle it there haha.*
RW: *In January, no less! Feb now, I guess… I'd like to find storage in Eugene, that would be most convenient.*
PE: *I wish Eugene was as friendly as everywhere I went in Canada. Theres not one town I passed in Canada were everyone wanted to help.*
RW: *A good friend of mine is a general contractor in Vancouver. Maybe we can line up some plumbing work and you can get back to Canada.*
PE: *That would be awesome. Theres not much work here. First day back at work yesterday in 5 weeks. The jys of being self employed.*
RW: *How adamant was Jeff about scrapping your bike? Did he give you a STRICT deadline, or is he somewhat flexible? Can we buy more time with him? Perhaps tell him that we can for sure work this out in Feb.*
PE: *He seemed strict about the 31st. He might be able to be sweet talked…. But then I haven't had much luck in the past haha.*
RW: *I'll send him an email or FB message tonight. I'll make contact with him and tell him we'll sort it soon. Worst case scenario, I'll send the guy $50 to buy some time until we can find another spot to store the bike.*
PE: *You had anything from Dave? To say whats he up to?*
RW: *Heard from him today. No leads. Asked him if he'd been in contact with Matt Ivey. Haven't heard back from Dave on that. Do you ever hear back from Matt Ivey?*
PE: *Matt has a place to store it in space to store it down in Redding. Hes just had a baby so currently doesnt have much free time to get it.*
RW: *It's nice he offered. Yes, moving the bike around is the tough part. Here's an idea: Dave C flies to Eugene, finds a cycling partner, pedals to Cabo, Mexico, then sails to Hawaii. Two expeditions in six weeks! ;)*

PE: *Sounds like a great plan* ☺
RW: *Just need to find storage in Cabo!*
PE: *Maybe someone else can keep cycling it ha. Find enough people it might end up home in England in a few years.*
PE: *Oh gosh I should try sleep…*
RW: *Okay, mate. Go get some rest. Chat again soon. Keep me updated. Cheers.*
PE: *Good talking to you, chat soon.*

Later that day, I sent an email to Jeff Lozar, the guy storing Paul's bike in Eugene, Oregon.

Hi Jeff,

My name is Rod Wellington. I live in Vancouver, BC. I am an acquaintance of Paul Everitt. I understand you are storing his bike car in Eugene. Paul has informed me that you have asked for a storage fee of $50/month starting February 1st. I also understand that you have been storing the bike since October or November 2011. From what Paul has told me, you are prepared to scrap the bike if the monthly storage is not paid. I know that Paul has offered the bike for sale and, as of yet, has not found a buyer. I also know that you have been storing the bike free of charge since last year. So, I see that both sides here are making an effort to accommodate the other. Obviously, you have come to a point where, if you are still going to continue to offer storage, you will have to charge for it. I can appreciate your position. And I can appreciate Paul's as well. He is England and is unable to physically deal with the bike. Now, please appreciate my position here. It is not my intention to step on anyone's toes. My intentions here are to ensure that the bike does not get scrapped, as we have future plans for it, and to ensure you that someone in North America is dealing with this issue.

Paul and I, along with an adventurer friend of ours, Dave Cornthwaite www.davecornthwaite.com (who is interested in using the bike car for an upcoming expedition), are all looking into alternate storage options for the bike car. Admittedly, we are looking for free storage options. Ideally, we are hoping to find someone in Eugene who is willing to store the bike for free. By finding someone local, it will make transporting the bike that much easier. We are also sourcing out other storage options in the Northwest and northern California. Our hindrance at this point is the inability to transport the bike to a location outside of Eugene. If we can find someone willing to store the bike, as well as pick it up in Eugene, that would be ideal.

I want you to know that we are working on this conundrum and we will find a solution. As I've said, we have plans for the bike and we are committed to deal with the storage issue. We do not want to see the bike scrapped or sold. We want to assure you that the bike has not been abandoned. We ask that you continue to store the bike for the time being and please rest assured that we will find a solution in the next couple weeks. If need be, I am prepared to travel to Eugene to deal with it personally.

I look forward to reading your thoughts on this matter. I can be reached via email or you can message me on Facebook (see links below). You can also reach me by phone in Vancouver at ***-***-****.

Rod Wellington

On February 2, 2012 at 2:27PM, Jeff Lozar wrote:

Rod,

Sounds like you have a good grasp of the situation with the bike car, I will add my thoughts to the discussion.

Paul walked into my office on November 2nd of last year in a somewhat desperate situation with having to leave his bike car behind. I felt bad for him and gladly found an out of the way place for it to stay. His story on that day was he had a buyer in place but she was unable to pick it up for a few days. The Distinct impression Paul left with me was this bike storage was a very short term thing. I have since posted a couple of reminders on his "going solo" facebook page and emails inquiring about getting a permanent home for Priscilla. Sometime in December I sent him an email informing him of the January 31st deadline. I told him I would not keep it past the 31st. With the 31st approaching and still no one to take the bike I came up with the paid storage option. I will not scrap the bike car, that would be a waste. As I see it I now own the bike car and it has a $50 price tag on it. If Paul wants to keep his options open he should send me $50 for the month of February at his earliest convenience.

Jeff Lozar

I finished reading Jeff's email and thought,
"Well, well, well. This looks like a standoff. Both sides are willing to be flexible, but neither side wants to give in. Looks like someone needs to step up, and I guess it's gonna be me."
Meanwhile, back in Memphis, the SUPAS train rolled on.

On February 1, 2012 at 12:03AM, Jonathan Brown wrote:

ALOHA!!!

DATE CHANGE: Sat. April 7th - Sat. April 14th
Dale and Robinson got together today to talk about ideas and realistic expectations. They concluded that there would be no way we could confidently promise an end date on the 14th if we start on the 9th.

The start date will have to be Saturday April 7th. It's the only way we'll be able to fully promise and meet expectations without pushing us to a possible breaking point.

I hope this doesn't change things for any of you. I'll need to see if I can get those additional days off myself. Please let me know this changes your scope of commitment.

TEAM CHANGE: Goodbye Harry & Hello Dave (Maybe) & John Hyde

NO MORE TEAM MEMBERS
Our team has grown quite a bit. We potentially have 9 team members. I was talking to Dale and both of think we need to stay with who we have from this point on. If you think of anyone that may want to paddle please promote the opportunity to paddle sections with us. This way our core team stays somewhat small and manageable and we'll be able to raise more money for the cause. Only guys that haven't been confirmed are Dave C. and Matt Farr.

JOHN HYDE
Just talked with John Hyde tonight. Wanted to get him included in our communication. He is one of few who have paddled from Black Jack Rd. to the Mississippi and all that done with his 15ft canoe! He'll be paddling with us from Baker's Pond to Old 72 but not sure after that. He wants to do a paddle in that area sooner than later so if you're wanting to explore before our paddle connect with him for sure. John is a great resource!

HARRY
Also, I am bummed to say that Harry Babb won't be able to paddle with us. He had been planning a trip through the states and was not able to move those dates around. He'll let us know if things change. Maybe someone could ensure his van "breaks down" before he goes…. JUST KIDDING HARRY!!! :)

DAVE CORNTHWAITE
Just when you thought Dave C. wasn't going to paddle with us… BOOM… WRC and friends pulls the rabbit out of the hat and found a way to pay for Dave's expenses. Now here's the catch… Dave needs to find a cheap way to change his ticket dates and re-route his flight to the Mighty Memphis! So, it's not confirmed but we'll let you know.

MEETING PLANS: Need to plan a meeting date and time.
This past Monday a lot of good things happened and has now launched Stand Up Against Slavery (SUPAS) into 5th gear. I'll be trying to set a date for the team paddlers, reps from OBS, WRC, BCCC, and

Outdoors Inc. to get together, meet, and discuss details concerning the SUPAS paddle.

Could you please let me know times in the evening or on a Sunday you could be available. I'll try my best to coordinate a time a place best for all.

TWO MORE COOL THINGS: This is pretty neat :)
1) Just heard that we may have Rainbow Sandals on as a sponsor.
2) A well known SUP company, C4 WATERMAN, has agreed to help by giving us pro-pricing for their boards. They have some great inflatable SUP boards that will be great for this trip. Check out www.c4waterman.com and look at the iTrekker. If you're interested in getting a SUP for this trip please let me know!!!!

Jonathan Brown

On February 1, 2012 at 12:03AM, Robinson Littrell wrote:

For those of you whom I haven't introduced you to, my name is Robinson (you can all me Robo for short) and I'm on the OBS end of things. Until Rachel comes back, I'm going to be the guy to talk to about sponsorships and donations and partnerships for the most part. I definitely want to thank WRC for their great work and thank them for joinging with OBS for this joint cause- to preserve the wolf and to put an end to sex slavery! Let's get rolling!

Soli Deo Gloria,

On February 2, 2012 at 9:16AM, John Hyde wrote:

I am John Henry Hyde - I am one of two people alive who has actually canoed from Blackjack Road - to the Cobblestones in Memphis. My phone number is ***-***-****.

One bit of input from me - anyone planning to do this trip on a paddleboard should modify the front and install some sort of pull-handle because with all the Beaverdams, and logjams - you will necessarily be pulling whatever watercraft you utilize over them, and through treetops! The bottom has lots of downed trees! Also - Lacrosse Burly boots, or some sort of waders, should be standard equipment for each paddler, as there will be MUCH opportunity to propel your watercraft from outside your watercraft! Again - this is the most snake-infested swamp I have ever been in (down at least as far as Michigan City), and the bottom is quite inhospitable! If you aren't familiar with Water Moccasins, you WILL become so before you see Michigan City (there is a GREAT little whitewater rapid just downstream of Michigan City.). :-)

On February 2, 2012 at 9:54AM, John Hyde wrote:

Today I drove from Memphis, re-connected w/my local contact 1/2 way between Blackjack Rd and New 72... And - I also went to Blackjack Rd, to Goose Creek Rd, and on to Baker's Pond... The river, coming out of Baker's Pond is like a 18-24" trail w/4-6 in inches of water running down the middle... Not sure how far it remains Un-floatable, but it will definitely require quite a bit of pulling whichever watercraft you choose to utilize. Be prepared to suffer through the most brutal bottomlands which the Southern USA has to offer! :-)

    I wasn't quite sure what to make of these emails from John Hyde. They struck me as the rantings of an old redneck who'd spent too much in a swamp, a gun-toting moonshiner who'd just as soon light up your rear with a shotgun blast if he caught you on his land.
    "Where the hell did they find *this* guy?" I thought out loud.
    His words seemed saturated with hyperbole and half-truths. I was on the verge of dismissing his fearful rants when I decided to re-read them.
    I reminded myself that I knew nothing about the upper Wolf River—its source, its course, its place names. Blackjack Road, New 72, Goose Creek Road, Baker's Pond—none of this sounded familiar. I knew nothing about travelling through snake-infested swamps. But, from the sounds of it, John Hyde did.

"Maybe I should give this guy the benefit of a doubt," I thought as I re-read the emails again.

This time, some of John Hyde's deep concern seeped into my brain. He knew he was dealing with a group of city slickers. He was letting us know in no uncertain terms that paddling the Wolf would be anything but a walk in the park.

Up to this point, I was comfortable embarking from the river's source with a GPS and a topographic map. As long as there was a river course to follow, and an increasing flow to match, we'd be fine. But now, having read the fear embedded in John Hyde's words, I was left doubting my own abilities, despite the fact that I'd been through some extremely challenging ordeals during my own expeditions.

Had I waded into this latest challenge without fully thinking it through? Had I said "Yes" too soon? Seriously, what did I know about snakes, swamps, and stand-up paddleboarding? A scared, little part of me wanted to shirk away for a short while and sort out these newly introduced fears. It needed a nail-biting session filled with worry and woe. It needed to know everything was going to be okay.

I forwarded John's emails to my sister, Carrie. She read them and quickly responded.

"Wow. Sounds dangerous," she replied. "I don't like snakes. Sounds like you better get a pair of water moccasins."

I had to inform her that water moccasins were snakes, not shoes.

Carrie's naivety wasn't unfounded. She lives in an area of Ontario where water snakes are rare and poisonous snakes are rarer. She, for all intents and purposes, was a city slicker (even though she lives in a rural area). Still, I found her uninformed knowledge of swamp travel humorous and the lightness of her response dissipated my worry.

"I mean really," I said to myself, "how hard could it possibly be?"

SUPAS – Day 3 – April 9, 2012 – Easter Monday

"I'm never going to forgive John for leading us into this god-for-saken hellhole!!!"

The screaming voice ricocheted around my brain like a bullet fired into a metal bottle. It burst forth from a place of pain, a place of intolerance, a place flooded with profanity—vulgarities that would offend every church-goer within earshot. I desperately needed to unleash an explosion of expletives as wide as the fucking swamp I was in, the swamp I desperately needed to exit before it voraciously consumed the last light of day. And judging by the grey gloom I was currently trudging through, that moment was set to happen soon, too soon for my liking.

Robinson's thin teenage frame stumbled through the dusky murk just yards in front of me. His actions seemed deeply lodged in fear, and his fear both hindered his progress and ignited his expediency. He wanted out as much as I did—maybe more so. He was shirtless, half-naked in a water-world bursting with lush vegetation and hopelessly maligned with a tortuous maze of fallen trees. It was not a place for humans, especially half-naked humans, especially half-naked, god-fearing humans who planned for a shorter day of less challenging adventure. And yet here was just such a specimen, trapped in the company of an impatient atheist who stubbornly refused to refer to himself as an atheist. The enraged unbeliever had also planned for a shorter day of less challenging adventure and was moments away from loudly voicing his displeasure, regardless of who heard it. For three full days he'd fiercely bitten his tongue when his anger threatened to boil over, fearful of how his paddling partners might view his outburst. He felt such a flare-up would cost him precious credibility among people he'd only recently met. He desperately wanted these people to like him, to embrace him as one of their own. He wanted to offend no one, yet he needed release, he needed to vent, he needed to speak his mind regardless of whether his words would upset his conservative counterparts. He had enrolled for this expedition with the distinct purpose of bolstering his persona as an "adventurer." He hoped the resultant exposure would net him additional book sales and lucrative speaking engagements. If he slipped up now and let loose a brazen tirade upon this lot of Bible-Belt-based paddlers, any hopes of reaping said benefits would surely be dashed. And so he plodded silently on, staring at the naked white back of a god-fearing teen who hurriedly towed a 12-foot plastic paddleboard over, under, and through a chaotic wooden web of fallen trees and upturned detritus.

Gone was the big, beautiful swamp, where, hours earlier, the atheist-who-refused-to-refer-to-himself-as-an-atheist spent considerable time sharing smiles with the god-fearing, shirtless teen, both of them admiring the exquisitely expansive sea of green lily pads that spread like a carpeted ocean to the forest-filled horizon. Now, they were entangled in a seemingly endless, hellish quagmire.

"Go!" I shouted in my head. "Keep fucking moving!!"

When I neared Robinson's naked back, I felt an almost uncontrollable urge to push him, to propel him out of the morass, to drive him from my vision. I needed out of this tortuous tangle of tree trunks and fallen limbs. I needed to hunker down unmoving in my tent and stare through darkness so black it hurt my eyes. I needed refuge. I needed this punishment to end.

John Hyde had warned me. He had warned us all.

"The last mile will be the toughest," he told us while we semi-circled around him in the mid-morning sunlight. John knew the route better than any of us and he strove to fearlessly lead us through—as he put it—"some of the most brutal bottomlands which the Southern USA has to offer."

"Y'all better save some energy for that last stretch," said John as he drained a generous mouthful of Guinness stout from a sweaty bottle during our lunch stop. "It's a *real* motherfucker."

John was right. It *was* a motherfucker, a *real* motherfucker.

I heard the deep gouges on the bottom of my plastic paddleboard catch and noisily scratch across the rough bark of the fallen tree I'd just hoisted myself over. A six-foot length of nylon rope tethered me to the paddleboard's bow. I'd been pulling that rope taut for the majority of the previous 72 hours. I could feel and hear the resistance as I dragged the board over the tree trunk. It teetered momentarily, weightless and noiseless. I yanked the board over the fulcrum and heard it slide against the bark and crash into the murky river. I quickly moved forward to the next fallen tree, and the tree after that. And the tree after that. Trees, water, and looming darkness filled my periphery. And smack dab in the middle of that periphery was Robinson's bare back, stark white against the gloom. There was no escaping the shirtless teen.

    John, Chris, and Phillip were well behind me, somewhere upstream. In an effort to reach our designated campsite before dark, Robinson and I had quickened our pace. In doing so, we'd significantly increased the distance between us and our paddling partners. I stopped for an instant and listened. For the longest time, I could hear their grunts as Chris and Phillip heaved Big Bertha (their 19-foot canoe) over the fallen trees. John would often stop after he'd wrestled his own canoe over the obstacles, and double back on foot to help them with Bertha. But now, in the stillness of that one resting moment, I heard nothing. Robinson had stopped as well. I could see his white shoulders rise and fall as he inhaled and exhaled deeply. He was waiting for me to move, waiting for a sound of movement from behind him. He would proceed when I did. He was waiting for me. He wouldn't abandon me. Nor would I abandon him. We were in this together and we were going to finish together. It was a comforting thought. After all, he was 18 years old—just a kid. I couldn't leave a kid alone in a swamp, especially at nightfall. I was committed to getting him safely to the highway bridge downstream. I would do that for him, for that church-going, god-fearing shirtless kid. I owed him that much. Somehow, he'd braved this damn swamp and made it to the day's end. He was a trooper, just as stupid as the rest of us for being out in this god-forsaken hellhole. Just as stupid as John, Chris, Phillip, Jonathan, and me. And Dale. Yes, as stupid as Dale.

    "*Dale!!!*"

    My silent shout ricocheted around my brain and my face scowled at the thought of his name. A flash of anger surged through me.

    "FUCK!!!!" I shouted.

    Robinson's head shot 'round as he spun to look in my direction.

    "You okay?" he shouted.

    "*No!*" I shouted back.

    Anger washed over me as I slung a wet leg over the rough bark of a fallen maple tree.

    "Fuck! FUCK!! FUUUCCCKKK!!!"

Rage coursed through my body and I felt my heart pounding loudly in my chest. I was livid. Robinson turned and plodded onward. Through squinted eyes I could see his bare back, a white swaying beacon amongst the silhouetted tree trunks. I followed it. We were braving this battle alone, just the two of us. The men who knew this river—John and Dale—were nowhere to be seen. We had abandoned them, maybe for own safety. Or maybe they had abandoned us. Whatever the case, I was livid. I felt the team should be conquering these obstacles together—no splintered sub-teams, no "going it alone." But Dale thought differently. He wanted a leader elected, someone who doled out decisions for the team. I believed he secretly wanted to be that leader. He wanted to take the lead and set the pace, and he wanted the rest of us to keep up with him. He'd ranted to me that very morning about how Jonathan Brown should be stepping up as the team leader. He even went as far as to pitch the demand to Jonathan. But Jonathan had craftily defused the situation by stating that he wanted the expedition's leadership to be a democratic process, one where the team made decisions based on feedback from all members, not just the one with the most experience. I sided with Jonathan on this subject. I didn't like the idea of handing over my decision-making power to Dale. Based on what I'd seen of Dale's behaviour in previous paddling excursions, as well as the first 48 hours of this expedition, he was a power-hungry dictator who believed his solutions were top shelf. I believed otherwise, and so did the rest of our team. Now, as the light of day faded and the forest closed in around our broken team of paddlers, Dale had gone AWOL. No one had seen him for hours, and no one seemed to care.

"Fuckin' *Dale*!" I seethed as I swung another wet leg over a half-submerged tree trunk. "What the *fuck* is with that guy?"

I felt the nylon rope go taut as I angrily hauled the paddleboard over another fallen tree.

"*Dale Sanders.* Mr. *Navy.* Mr. *River Guide.* Mr. *Wolf River Conservancy.* Mr. *I'm In Fucking Charge.*" The words left my mouth laced with derision. "He who wanted leadership has now gone fucking *AWOL.* Well, fuck me with a fallen tree. Doesn't that just top everything?"

I shook my head at the absurdity of it all. Deep down, I hated him for abandoning us. I was glad he was gone. Deeper still, in some soft recess of my head and heart, I was worried about him. Had he reached the camp at New 72 (the Highway 72 bridge that spanned the Wolf)? Was he lost in the swamp? Did he need our help? Was he dead?

Finally, but not surprisingly, my tempered resolve snapped like a twig in a hurricane. No longer could I hold in the rage.

"Where the *FUCK* is Dale?!?" I screamed. "FUUUCCKKKK!!!"

My voice echoed through the forest and its booming volume shocked me. Robinson stopped, turned, and looked at me. I met his gaze. He asked no question this time. He simply turned and plodded onward in the gloom, his bare back barely visible amongst the growing shadows.

The following is a mass email sent out by Dale Sanders.

On February 3, 2012 at 7:50AM, Dale Sanders wrote:

On Easter weekend, 10 brave souls, whose names appear in the "To" column above, plan to start a journey, paddling the full 105 mile length of the Wolf River. Sometimes on Easter weekend, most will launch from Bakers Pond (BP) area* Some will SUP (Stand Up Paddle), others will use canoes. Because of Easter, couple paddlers will meet the original crew on Monday morning, somewhere south-east of Mississippi Highway 72. All paddlers plan on paddling into Mud Island (MI) mid afternoon Saturday14th. Unfortunately, Harry Babb may not be able to join us and Mat Farr still has not confirmed.

I could not be more excited, more proud for our community, than I am working with you folks on this Operation Broken Silence's (OBS) paddle event called "Stand Up Against Slavery" (SUPAS). We are most fortunate Jonathan and OBS have voluntarily taken on this whale of an event.

I sincerely express my appreciation to all of you, and especially to Dave

RIVER ANGELS

Cornthwaite and Rod Wellington for their love for the Memphis area, their willingness to put aside personal plans, travel here and SUT/Paddle the Wolf, in this event, along side many of us local folks.

What these two guys have to bring, is monumental, in terms of good will, education and the promotion of our Memphis area causes. We must not forget, both Dave and Rod, themselves, represent great achievements and internationally recognized programs. These gents are truly remarkable and I challenge each of you, to learn as much about Dave and Rod as possible, prior to their arrival: http://www.davecornthwaite.com/ and http://www.zeroemissionsexpeditions.com. (N.B. Website now defunct. New website: www.rodwellington.com.)

OBS and the Wolf River Conservancy (WRC) are blessed to have volunteers, such as Dave, Rod and your selves, helping promote this/our historical and difficult upcoming paddle excursion.

My intentions in this Email are not to make extra work for anyone. Please, you don't even have to respond here, unless compelled to do so. After-all we are all volunteers. The only thing I ask, is for the paddlers and those info copy to individuals, to help us promote each cause: Dave's, Rod's, OBS, WRC, Bluff City Canoe Club (BCCC) and sponsors yet to be named. Their stories needs to be told. I don't believe any group has ever taken on such a difficult, un-known to the paddles, venture down this Mid-South's fame historical "Pirates Haven", "Blackbird River" and modern day "Wolf River".

The lower Wolf is well defined, mapped, traveled and understood. As matter of fact go to http://www.woplfriver.org (www.wolfriver.org) *and click on "River map" and you will see the best lower Wolf access map available, again created by one of our volunteers Mike Watson of WRC and BCCC.*

The upper Wolf is something else. I know virtually nothing about the sections between BP (Baker's Pond) to New Highway 72 in the State of Mississippi. To my knowledge there are no good maps that follow the actual river bed, where one exist, and swamp navigation. Google Earth probably shows the best pictorial view of the area. Also John Hyde does have some other maps he is working with in an effort to determine the upper Wolf River navigable waters and possibly the true source of the Wolf itself. I have heard, through Keith Kirkland, that one individual, maybe traveling alone, allegedly paddled the entire length of the Wolf. Close behind, I believe, is John, whom we are fortunate to have join us on this expedition. John has single handily paddled most all sections of the Wolf. The details John will, from time-to-time, pass on to the paddlers, could determine project success of failure.

I am most concerned about the estimated number of days it will take us to paddle, walk and crawl from BP (Baker's Pond) area to HWY 72. I do know BP is a well traveled and hiked area, with easy access and fairly well known. One can reach BP just off Mississippi, Highway 72, south only a short distance. The swamp areas Southeast of New Highway 72 are massive, seemingly un-navigable. From Highway 72 on to MI (Mississippi River), I have paddled before, and have stories to tell - we actually actually lost the river twice. From Michigan City, there are only two sections that will be a little difficult but, with experienced paddlers, these sections shouldn't be a problem: (Michigan City to LaGrange and Collierville Arlington to Walnut Grove Road). The section between Moscow and Rosville, TN will take little longer than expected because of numerous tree blockage, requiring portage.

I believe the number of paddlers that can be accommodate, for the full trip is maxed out. However, those of you, that would like to experience this adventure, paddle a section(s) with us, invitations will be forthcoming. Section paddling will be open to anyone pledging sponsorship support. Details will be released at a later date. If you know of anyone that may be interested please feel free to forward thisEmail on to them or let me know their Email address so I can keep them informed.

* John Hyde and others, including my self, are still trying to locate upper

Wolf River navigable waters. Due to sea level elevations and other factors, Bakers Pond might not be the true source of the Wolf River. Hopefully, primarily through John's efforts, we will have a better understanding of the springs that feed the Wolf River. For now, any reference to Bakers Pond (BP) is for general launch area Identification purposes only.

Very Respectfully, F. Dale Sanders

On February 3, 2012 at 10:36PM, Jonathan Brown wrote:

There's been a lot happening as you may have realized by looking at your inbox this week.

Just wanted to bring it all together in one place to ensure everyone's on the same page.

FINAL DATE FOR THE PADDLE:

START: Saturday April, 7th
END: Saturday April, 14th
* Reason for the date change: After looking realistically at the paddle it was concluded that we would not be able to meet certain deadlines if we started any later than April 7th.
* If you have any issues with these new dates please let us know. I think I've heard from everyone on the team and sounds like nobody has a problem except for Richard, who is ok with meeting us on Day 2.

PADDLE TEAM UPDATE:

Good News!!!
Dave is officially on-board… literally!!! Thanks to WRC and the generous help from another Dave's flight, food, and housing will be fully covered! He will arrive Friday April, 6th and join us from beginning to end. After the paddle he will spend the next week with WRC speaking at various schools and gatherings.

Bad News!!!
Harry Babb will be unable to paddle with us. He had prior plans to road trip to the West Coast and was unable to change his dates to accommodate for our paddle. Harry!!! You will be sorely missed! We hope that in your stead your dad may be able to offer help and carry on Ghost River Outfitters representation.

Just News…
We are only waiting to confirm one more paddler. Matt Farr was recommended by Adam. He is an experienced paddler and works with Shelby Farms. We'll keep you in the loop!

Team Summary:
As of now we have 7 confirmed Team Paddlers. Possibly 8 if Matt joins. We are closing any more entries onto the core team.
Please begin to promote the opportunity to join us on a donation basis per section.

SPONSOR UPDATE:

Thanks to Dave we now have Aquapac as an official gear sponsor. All confirmed Team Paddlers will receive an email with instructions about ordering your gear.

If you have connections for any gear, financial, or other sponsorships please let Rachel & Robinson know. Warm up the connection first. :)

UPCOMING MEETINGS:

WRC & OBS Meeting:
Date needs to be set some time after Wednesday, February 8th. Rachel will be back in town Monday Feb, 7th. Please connect with each other to nail down a time and day. I'm available most evenings and Fridays.

OBS & SUPAS Paddle Team Meeting:
As soon as we confirm our final paddler, I think it would be a great idea

to get together and if possible find some way to Skype with Rod and Dave. Any thoughts? Please let me know what days and times work best for you all.

The Event Name:
Stand-up Against Slavery (SUPAS)
The Sub-Title:
Stirring the waters of injustice

I'd like to ask, if you're not familiar with Operation Broken Silence's mission, purpose, and plans, please check out their website at www.operationbrokensilence.org and don't hesitate to ask Rachel and Robinson questions.

Here's their Mission Statement:

"Operation Broken Silence is a nonprofit that strives to protect the innocent by confronting social affliction and building a new generation of abolitionists who envision a world without mass atrocities and modern slavery."

Here's their current project:

Operation Broken Silence has begun taking steps to implement an aftercare facility for survivors of human trafficking in Memphis, Tennessee. A critical element of a successful aftercare system is the community, so we are launching CARE – the Community Aftercare & Restoration Endeavor.

I'll admit that the whole "human trafficking" thing was beyond the usual boundaries I set when associating myself with charities and fundraisers. Typically, I lean toward less politically minded organizations like food banks or local safety programs for schoolchildren. I've found those kinds organizations typically have less guilt-related weight attached to them. It's hard for me to articulate the feeling. It has something to do with the *seriousness* of the cause. For example, I don't attend war-related or civil rights-related demonstrations because the subjects of those causes bring me down emotionally. They don't resonate positively with me. Sure, I support the civil rights movement and I'm opposed to the military in all regards, but I'm simply not passionate about either topic. I'm not particularly passionate about food distribution for the needy either, but it is certainly more relatable to a wider swath of the public than many politically minded causes.

Being involved with Operation Broken Silence, even in a minor way, was oft-putting. I couldn't get fired up about the injustices being perpetrated on others, not enough to lend my full support to their eradication. Sure, I was opposed to such injustices, and yes, I supported the need for an aftercare facility, but truth be told, I wouldn't normally associate with a group like this. The content was "heavy," and I was not attracted to the drama surrounding the topic of human trafficking. I did, however, spend some time researching the topic and educated myself, if only marginally, to what was transpiring in the Memphis area. I'll admit I was shocked. I didn't realize human trafficking and sex slavery were widespread problems, not only in the U.S., but Canada as well. OBS was certainly not asking me to be a spokesperson for the organization, so my role in the fundraiser would be that of expedition team member. I would be paddling and spreading word of the event through social media. That was a minor role as far as I was concerned. I was comfortable with that. (To their credit, Operation Broken Silence was avidly campaigning against the genocide in Syria in early 2012. Three years later, the fallout from that genocide, and the mass exodus of refugees from Syria would became headline news around the world. My introduction to the topic came via OBS.)

Meanwhile, on the Bikecar front, a temporary solution had been found. I messaged Bikecar builder, Paul Everitt, via Facebook to let him know.

RW: *Hi Paul, Bike car update. I've been in contact with Jeff Lozar in Eugene. I am assuming ownership of the bike. You and Dave C will continue to have full access to the bike. Today I will be mailing Jeff $100USD, $50 to cover the cost of the bike, $50 for February's storage* *fee. The bike will stay at Jeff's until we find alternate storage. I have been contacted by a gentleman in Eugene who seems interested in storing the bike for us. I have left him a voicemail and am awaiting a reply. I'm hoping to have the storage situation resolved by February 11. Jeff Lozar has stated that the bike will not be scrapped. He now sees himself as the bike's owner, which I find odd. I still consider the bike as your property. Since Jeff has been trying to sell it for $50, I decided that I would "buy" it from him so that we could retain possession of it. If anything, the money is my way of saying "thank you" for storing it for these past few months. I will update you when I hear from the other guy in Eugene regarding storage.*

PE: *Cool beans, I'm glad the bike will be going to a good owner. Thank you for everything and I hope the Bikecar brings you much enjoyment as it did me. You should only need to pay Jeff $50 and not $100. The $50 is only for February's storage. He doesn't own it. I'm happy to give you the bike for free. Anyways thanks again.*

RW: *No worries, mate. I'll sort things out with Jeff. Do you have any plans to use the bike this summer?*

PE: *Nope, I'm just going to be floating down the Mississippi. Rod I would like you to be the new owner. I believe shes meant for you. She needs a new adventure. Get Dave to cycle it up to you haha. I believe that was his plans anyways.*

RW: *Sounds good. When does the Mississippi trip begin?*

PE: *If all goes to plan rear end off July. When I have finished that do you want a raft too haha.*

RW: *lol. Do you have river charts?*

PE: *Have not charts. Need to get some. Theres a lot of things I need to do but my day job seems to taking over my life haha.*

RW: *I went down the river from source to sea in 2001. I can send you a link for river charts.*

PE: *I would like that.*

Meanwhile, in Memphis, the members of SUPAS were busy talking about another river: the Wolf.

Dale and his friends at the Wolf River Conservancy had lined up a speaking engagement for Dave Cornthwaite at St. George's, an exclusive private school located on the banks of the Wolf in Germantown, a Memphis suburb. The expedition's schedule now revolved around this school event, creating an imposing deadline that would come back to bite us later. For now, we were all excited to bring SUPAS into the classroom with ol' Corn leading the charge.

Three members of my immediate family—my father (Robert), sister (Carrie), and brother-in-law (Mario)—began making plans to travel to Memphis for the SUPAS after-party on April 14. Dale Sanders was quick to offer his home as accommodation for my family—a kind invitation that they graciously accepted.

John Hyde had stacked up long hours scouting the upper Wolf and had made an amazing discovery about the river's source that would challenge not only the Wolf River Conservancy's notion of where the river began, but also his fellow SUPAS team members' beliefs. Unknowingly (or perhaps *knowingly*), John had unearthed a truth that would eventually drive a wedge between him and everyone else associated with the project.

On February 4, 2012 at 9:38PM, Dale Sanders wrote:

Paddled Houston Levee (HL) to Walnut Grove (WG) today. Took us 2.5 hours, even with high water. No way kids will paddle as fast as we. Bottom line, if we are to pick up students a St Georges, in order to reach WG, with time left to set up camp before dark, we will need to depart the school at NLT 1430. We should plan on arriving at the school NLT 1400. Is that doable?

Spoke several times with John Hyde last night and today. He seriously questions Bakers Pond (BP) as Wolf source. In short, he found three possible source (springs/lakes); at least one of which is at a higher elevation than BP. (armed with altimeter "aps" on his smart phone). Fortunately, the slightly higher elevation lake/springs area has more flowing water and better road access. (I hear the highest elevation spring is technically a rivers source). Sounds like a better launch site to me. You and I (plus others if they wish) need to meet John out there, possibly week or two before we paddle, if for nothing else, we will need a dry run before the Saturday launch. My vote is - what ever John recommends, for I know nothing about that area.

Looks like both Dave and Rod will be "camping out" in my basement, for the duration. Rod will be arriving on 4th, speaking to the (Bluff City) Canoe Club that evening. All you guys are invited. More details to follow. Dave will be arriving on 6th. Dave will be departing Memphis on Tuesday 17th. Rod may be staying couple days longer in the Memphis area. (His father, sister and brother-in-law will also be here). Dave is pretty well booked by WRC. Rod will have time to speak to school groups etc. basically anytime between Sunday afternoon - Tuesday or Wednesday, after the paddle. If you guys wish to offer his services, I can coordinate the request. Will need to get Rod's blessings prior to commitment.

For Jonathan. John and I would like to paddle from New Highway 72 to Michigan City, Mississippi, possibly this weekend or next week. What does your schedule look like? You could paddle with John in his boat; I will paddle my solo. This could be an important eye opener.

For all. I have a list of over 2100 active paddlers. We can use it it recruit section paddlers. (don't have Memphis White-water folks Emails yet).

Dale

On February 4, 2012 at 9:59PM, Dale Sanders wrote:

Old John is certainly a valuable team member. Some people are not going to like the facts that John and I have only begun to discover. We don't care who gets their noses bent out of shape, the true source of the Southwest Tennessee, Wolf River needs to be known. From here on in, the journey should prove interesting. (Hope we don't accidentally locate a Whisky steel or something). Seriously though, you guys will be shocked as to how remote and swampy the area actually is.

Ray (Graham), You might wan't to phone John. Believe he can paddle with you, at least part of your March upper Wolf trip.

Begin forwarded message:

From: John Hyde
Date: February 4, 2012 8:29:24PM CST
To: Dale Sanders
Subject: Advance Scout

I have located/proven the TRUE headwater to the Wolf River today...

Baker's Pond is NOT the highest in elevation (according to altimeter readings), OR in CFM (cubic feet per minute) (according to my eyes) as far as the three converging watersheds which make up the Wolf River Channel go...

There is Baker's Pond from the West, Goose Nest Farm Creek in the middle, and Booker's Creek from the East which all meet @ "The Old Brown Hotel" - as told to me by a member of the family who owns it. (This was the location of his Grandfather's deerstand) The locals claim this location as the Headwaters of the Wolf River. I - DO NOT!

The springs feeding the VERY PRIVATE lake on Goose Nest Farm appear (according to altimeter readings, and flow rate) to be the TRUE headwaters of the Wolf River. There is a somewhat-navigable waterway from which a "Source - To The Mississippi" expedition can be mounted just below this lake on Holly Springs National Forest land, and access is no big problem. It will require a fairly short portage, and is FAR superior to the one offered by Baker's Pond! It also offers a much-more-navigable waterway trailing out of it!

I have a local connection around 1/3-1/2 way from Blackjack to Highway 72 - and it may be possible to replenish supplies/gear from his place. I am still ironing out the details. Locals do NOT like outsiders in this area...

I will be glad to guide your expedition at least as far as Blackjack Road... I might continue on as far as New Hwy 72, as that is/was one of my favorite and most memorable runs upon the Wolf. I have done the rest of the Wolf downstream of Blackjack Rd, and any further accompaniment by me will be purely up to how I feel when I reach the "Three Bridges Road" - as the locals call it...

Sincerely,

John Henry Hyde

In 1998, Memphis paddlers Gary Bridgman and Bill FitzGerald became the first two people to descend the Wolf River from source to mouth. A story about their week-long journey was featured in the Commercial Appeal, a Memphis weekly entertainment newspaper. Bill and Gary (a Wolf River Conservancy board member) began their trek from Baker's Pond in northern Mississippi. They walked the first 12 miles of the river (Baker's Pond to Pipeline #5), and then paddled the remaining 90 miles to the Mississippi River in Memphis. Friend and fellow paddler, Larry Smith (former Executive Director of the Wolf River Conservancy), joined them during the hiking portion of their Wolf River adventure. Larry and Gary chimed in via email with words of wisdom (and caution) for the SUPAS team. Their stories, included below, are best told in their own words.

On February 5, 2012 at 8:40PM, Larry Smith wrote:

Hi all, better think twice before you do anything beside hike between BP and Hwy 72. No channel to speak of mostly braided channels and lots of marsh. Gary Bridgman did this entire trip in 97, he and I et al walked from BP to a point about 2 + miles above OLD Hwy 72. Also MS law says Wolf River from new Hwy 72 down is a public right of way, above Hwy 72 it is not, thus you are trespassing. Gary Bridgman got permission from all landowners to cross their property before we did so. Talk to Gary Bridgman he knows all about this, it was a long haul and a lot of work just for two people. It made the front pages of the Commercial Appeal, WRC had a huge newsletter story on it etc.

Larry

On February 5, 2012 at 10:48PM, Gary Bridgman wrote:

Larry is right, although I never like to discourage an adventure on the Wolf...and you may have some of these issues worked out already.

In helping Bill FitzGerald and I prep for the 1998 trip, Larry went to the Benton Co. courthouse (Ashland, MS) and made a copy of the property maps along the Wolf from Baker's Pond to Hwy 72, but they only listed names, sometimes as abbreviations.

Since it was 1998, and most people still had landlines (and white pages listings) I was able to search an online whitepages database for matching names in Canaan, Blackjack, Ashland, etc. and hit them with a quick, neighborly-sounding spiel about my trip, sticking with property on the right descending bank. When I couldn't find a number, I would ask the people whose numbers I COULD find if they had their neighbor's number. Sometimes they knew the number from memory, sometimes they wouldn't tell me....then the mystery neighbor would call me himself a day later, once the first neighbor had a chance to run it by him in confidence. We walked the right descending bank from Baker's Pond for about 18 river miles, sometimes having to walk a natural-gas-line right-of-way a few hundred yards upridge from the water. When we descended back to the "riverside" it was to a swamp where Tubby and Blind Tiger creeks meet the Wolf, and we had to do the equivalent of wading/log-hopping across a corner of the Ghost River swamp to get to where our canoe and tent were waiting. You will want an updated copy of that map.

Watch out for the Sparks lands, on the left descending bank a half mile upstream of Old 72. They run high-priced hunting outings—in season—and they don't want paddlers scaring off the turkeys. The Sparks' private game warden/naturalist there heard about my plans and told me on the phone, politely but firmly, that if he could keep me off the river there, he would. (Based on what Larry just wrote, perhaps he could have.) I held my tongue, not telling him what I thought about their attempts at creating a faux English countryside for their shooting parties by cutting trees all the way to the bank, which would likely result in erosion and channel shifts. But I knew the river would "straighten" that out on its own.

Be prepared to hoist your canoe over hundreds of obstacles while navigating a terrain worthy of military survival/special ops schools. I cursed all the gear I had piled into our boat that day.

Even with the help of the now-deceased and mourned Bill Lawrence, who knew that section better than anyone else, we spent long stretches scratching our heads over the useless topo map. (I once paddled my way down a waterpath, hoping to regain a real channel and found myself in a meadow. In retracing my path, I thought the "waterpath" looked suspiciously like a tractor's tire track.)

Here is my expanded journal/memoir of the trip, with photos. www.flickr.com/photos/wolfriver/sets/72157594572079868

-best
Gary

(N.B. An account of Gary Bridgman's 1998 descent can be found at the Flickr URL above. The text below is from the same Flickr URL. Text reprinted with permission from Gary Bridgman.)

In April 1998, the Wolf River Conservancy and Outdoors, Inc., helped me undertake a full descent of the Wolf River. My friend, Bill FitzGerald, and I hiked from the source spring in the Holly Springs National Forest to what was then considered the head of navigation and dragged a laden canoe over about 20 million fallen trees before violent thunderstorms improved our navigable water supply. We made it to the Mississippi River (at downtown Memphis) in about a week.

The photos here include shots taken during the actual trip and during scouting trips earlier in 1998 when we were trying to figure out the head of navigation. They appear in geographical order (going downstream).

We had different people join us along the way, including (in chronological order) Bill Lawrence, John Huffman, Naomi Van Tol, Larry Smith, Damon Lapinski, Ray Skinner, Mike Jones, John Farmer, Chris Stahl (who helped me plan the trip), Mayor Sharon Goldsworthy (Germantown), Bill Currato, Jane Ellen & Mike Rawdon, 2 cops who woke us up in Kennedy Park, the jerks who stole one of our stashed coolers, Councilman John Vergos, Director Cindy Buchanan (Parks Services), my mom and dad, and a few hundred other canoers and kayakers who unknowingly accompanied us as we completed the journey via our participation in the 1998 Great Mississippi Canoe & Kayak Race.

Here is my account of this canoe trip, published in Oxford Town and the Wolf River Conservancy's newsletter in the summer of 1998.

(N.B. Text reprinted with permission from Gary Bridgman, Oxford Town magazine, and the Wolf River Conservancy.)

A River Creeps Through It

by Gary Bridgman

(OT editor's note: On May 1, 1998, Ole Miss graduate student, William "Fitz" FitzGerald, became the first person in recorded history to travel the entire length of the Wolf River. WRC board member and Oxford, MS, resident, Gary Bridgman, became the second person to do this...about three seconds later (he was in the back of the canoe), as the two completed the "Wolf River Survey." Gary and Fitz hiked and paddled from Baker's Pond to the foot of Union Avenue to help raise awareness about the river

RIVER ANGELS

as a whole. Sponsors included the Wolf River Conservancy, Outdoors Inc., Ghost River Canoe Rentals, and BellSouth Mobility. What follows is Gary's rather unscientific, non-chronological account of the trip.)

There's a distinction between being drunk on a river and being drunk with a river. One does not need alcohol or drugs to have mind altering (or life changing) experiences in a canoe. Fast moving streams like the Nantahala and the Ocoee are what I call "adrenaline rivers," while the Wolf is an "endorphin river." It offers canoeists a priceless glimpse of what all other rivers' headwaters in this region looked like before the Corps of Engineers channelized them.

William Faulkner described such swampy, untamed rivers as "the thick, slow, black, unsunned streams almost without current, which once each year ceased to flow at all and then reversed, spreading, drowning the rich land and subsiding again, leaving it still richer." They are intoxicating, to say the least.

The Wolf River is teeming with wildlife and wetland vegetation, but my favorite part about our recent "expedition" was not its biodiversity, but its psychodiversity: all the interesting people I met in the process —- interesting people like the two cops who almost busted us for vagrancy.

"Good Cop/Bad Cop"

Memphis, May 1, 8 miles from the Mississippi River.

"Hey! Get up! MPD!" shouts a Memphis police officer.

William FitzGerald ("Fitz") and I are stumbling out of the tent into the glare of their Mag-Lites, my left leg is still tangled in my sleeping bag.

"What are you doing here?" the other officer calmly asks.

It's 3 a.m. We are camped illegally in a city park located on the Wolf, having built an equally illegal campfire. I've explained that we aren't vagrants and that there is a canoe hidden in the tall grass over there and that we're paddling the entire length of this river on behalf of the Wolf River Conservancy.

Now the policemen are more relaxed. They're even giving me pointers on how to delay being raped or murdered in case some of the local toughs come by. (It didn't look like a rough neighborhood from the river.)

We had been at it for six days by the time the police woke us up in Kennedy Park: hiking and paddling (and wading) some 90 miles by that point. Just a few more miles to go to reach the Mississippi River.

"Thirteen Weeks Earlier"

Moscow, Tenn., January 24

The whole thing started when my friend Chris Stahl, who runs a canoe rental service on the Wolf River, asked me how he could attract more people to the river. "Canoe the whole thing in one lick, man," I said, not very helpfully.

Chris was asking me for ideas about popular day trips for families and church groups, not about some kind of pilgrimage out of the heart of darkness into the middle of industrial North Memphis. There were remote sections of that river no one had navigated in decades —- too shallow, too narrow, too overgrown, too full of fallen trees. We could count on crawling out of the canoe to lift it over logs several hundred times in the process.

Chris liked my thinking anyhow, but business commitments and common sense kept him on the shore for most of the trip. So I enlisted Fitz to make the trip with me instead. From January onward, one or both of us spent nearly every weekend scouting different sections of the river and meeting peculiar people.

Walnut, Miss., February 8

"You can put this in the Bible if you want to, but I like snakes more than I like most people," said one man we met while scouting a swamp. "You can trust a cottonmouth; all you have to do is know how his mind works." He viewed our "People's Republic of Oxford" Lafayette County license tags with suspicion, wondering if we were more "dope smoking a__holes" trespassing on his land, but we've since developed an interesting friendship.

"Gary, so far I think you're a decent person, but if you ever cross me, I can give away one of my motorcycles to someone in Memphis who'll do anything to you that I ask!" Great. I gave up being a Republican for this?

"The Trip Begins"

Baker's Pond, Holly Springs National Forest, April 25, 98 miles from the Mississippi River

We had to hike around and wade through 18 miles of swampy bottomland this first day of the actual trip. (Our canoes were waiting for us downstream).

When we scrambled up to the first dirt road that crossed the Wolf, a nice lady in curlers skidded her old pickup truck to a halt beside us. "Are y'all the canoe people?" she asked with a disbelieving smile. We were now 30 seconds into our 15 minutes of fame.

Canaan, Mississippi, April 26, 80 miles from the Mississippi River

This was the hardest day of canoeing in my short life. Fitz and I were joined by Ray Skinner (pictured above) and Bill Lawrence, who is something of a Yoda or Ben Kenobe figure in the uppermost Wolf and an invaluable guide to us for this section. We pulled our gear-heavy canoe out of the shallow water and over fallen trees almost every 150 feet of river channel. We only made five miles that day. It rained its butt off that night, which was good. Come Hell or high water, I'll take the latter.

"More Cops, Three Mayors, and a Waitress"

LaGrange, Tenn., April 27, 60 miles from the Mississippi River

I was driven up to town from the river bottom by a Fayette County sheriff's deputy at the end of a long, but very productive day —- triple the mileage of the day before. The deputy had been dispatched at the request of Mayor John Huffman of nearby Piperton, Tennessee.

John, who is also the president of the Wolf River Conservancy, was having a lot of fun keeping track of us via walkie-talkies. Here's an excerpt from and e-mail he copied to dozens of people two hours later: "Who would like to bet that this was the only time in young Bridgman's life that he was happy to find out that the Law was looking for him? With the lightning and heavy rain present in Fayette County, they are no doubt thinking about how it might of been if they had not made it to LaGrange and been forced to camp along the river."

Actually —- at that very moment —- I was thinking about pouring another glass of cabernet while that massive thunderstorm was making the lights flicker. Fitz and I were holed up in a bed & breakfast two miles upland, owned by a Conservancy member. I refilled the glass of LaGrange's mayor, Lucy Cogbill, who stopped by to check on us and enjoy a dry view of the passing monsoon from the back porch.

But I was also thinking about how the mayor of Rossville, Tennessee (25 miles downstream) didn't give a crap about our expedition because he was having to supervise the partial evacuation of his town due to flash flooding.

My friend Naomi visited briefly, then drove west back into Memphis along the length of the river's floodplain. "Driving out of LaGrange," Naomi wrote in her own mass e-mail report, "the radio was reporting: flood advisories for Collierville; tornadoes in northern Mississippi; and flash flooding, evacuations, and possible road closure at Rossville. This should make for a speedy and exhilarating ride for Gary and Fitz tomorrow."

RIVER ANGELS

Rossville, Tenn., April 28, 45 miles from the Mississippi River

Exhilarating. Right. More like "intimidating," as we constantly ducked under tree limbs that were coming at us at twice their normal speed. I took the only unplanned swim of the trip after being swept out of the canoe by one of those passing limbs.

Fitz is a very even-tempered First Lieutenant in the National Guard, but he sounded more like a drill sergeant as he coached me up onto a half-submerged tree. "Get up on that tree, Bridgman! Let's get some adrenaline flowing!" he shouted. I obeyed both commands. Fitz carefully maneuvered the canoe underneath my unsteady perch, enabling me to flop down into the boat like a stunned raccoon.

That night, near Rossville, we stayed in a hotel after stuffing ourselves at the Wolf River Cafe. Our waitress, Dorene, was the first of many people to give us the once-over, trying to figure out why we were wearing two-way radios and carrying cell phones while our shabby personal appearance suggested that we lived in an abandoned station wagon.

Earlier that morning, Fitz and I floated through the most amazing stretch of the river, known popularly as the Ghost River section.

Keith Kirkland once described it this way: "About halfway through the trip, small braids of river begin to split off the main channel, disappearing into a dense, standing-water Cypress-Tupelo Gum swamp just before the river abruptly hits a dead end. Only one among the dozens of narrow, twisting corridors splitting off to the left of your canoe will lead you through the full mile of swamp. The rest dissolve into a forest of impassable knees and floating islands of Itea and Buttonbush. The river seems to be everywhere, but nowhere - like a disorienting funhouse hall of mirrors."

April 28 was my 35th float through the Ghost River section and in our haste we paddled it in near-record time, but it's never, ever a "routine" trip for me. I see something new and wonderful every time!

Germantown, Tenn., April 29, 15 miles from the Mississippi

The next mayor on our itinerary was Germantown's Sharon Goldsworthy, who fed us her prized beef stew and corn muffins while hearing about our progress.

The next cop on our itinerary was at Germantown Centre, the city's sprawling performing arts and recreation complex.

"Hello, Mayor!" he said in a cheerful-yet-bewildered tone as Sharon walked us through the health club on the way to the showers. It was fun watching his eyes dart back and forth between his commander-in-chief and the two muddy hoboes trailing her.

"The Voyage Home"

Memphis, May 1, 0.5 miles from the Mississippi

The journey began where the Wolf River is three feet wide, in a county that hasn't a single traffic light. On this last day, in the shadow of the Pyramid, it was nearly 300 yards wide.

I was glad to see that Wood Ducks and Great Blue Heron were thriving on the river all the way downtown.

As we passed under the Hernando DeSoto Bridge (which also spans the Mississippi) and then the monorail bridge leading to Mud Island, within sight of the mouth of the river, we heard a terrible racket: screaming school children.

"Two, four, six, eight, who do we appreciate? Gary and Fitz! Yeahhhhh!" they chanted, having been tipped off about us earlier.

This "endorphin river" was becoming more of a hallucinogenic river. Speaking of which . . .

The night after my first float through the Ghost River section, in 1992, I had a weird dream. No plot to it really, just an image of the water slowly flowing in the darkness, beneath the canopy of trees and dense shrub and rotten logs, while I lay safe in my Midtown Memphis home.

I remember feeling strangely guilty that I wasn't still out there with the current, but also relieved to no longer be in that stygian gloom. I've since come to love that gloom, and all the surrounding light that defines it. And as Fitz and I neared the Mississippi River, I knew that I had finally accompanied that current all the way to its home.

Gary Bridgman is a WRC board member whose devotion to the Wolf River's protection is only equalled by his penchant for getting gloriously lost in its swamps.

Copyright 1998, Oxford Town, Wolf River Conservancy, Gary Bridgman

---

Bridgman's Flickr page (see link below) featured several photos taken during his 1998 descent of the Wolf River. The accompanying photo captions gave further insight into his journey down the Wolf. (Some of the captions appear below.) Bridgman regularly credits Memphis paddler, Bill Lawrence, with making the descent a success. Lawrence passed away in 2011, but his extensive knowledge of the upper Wolf River lives on in people like Gary Bridgman and SUPAS' John Hyde. Bridgman and Hyde often refer to Lawrence as "Yoda," a generous nod to the revered Jedi Master character in the Star Wars movie franchise.
(www.flickr.com/photos/wolfriver/sets/72157594572079868)

Caption from photo of Bill Lawrence upstream of "first bridge" across the Wolf, two miles downstream of source (Baker's Pond):

"*Bill Lawrence was our wizard for this trip. He normally explores the Wolf's uppermost headwaters alone, typically putting in near a bridge and paddling upstream until he's ready to go home. He was initially described as a guy who's hard to track down or meet. But I called the number someone at Outdoors Inc gave me and left a message. He called back within 5 minutes and within 30 minutes, Fitz and I were driving north from Oxford while he was driving east from Memphis. We scouted Baker's Pond for the first time and then we located the "first bridge" just above the Goose Creek confluence and paddled around in this beaver swamp.*

He joined us for several sections of our full descent of the Wolf, providing us with much needed navigational help on the first day in the canoe (day one of the trip was on foot).

Funny. As well as I got to know him, the warning that he's hard to track down was accurate. I live in the same city, but I haven't a clue as to how to reach him now in 2009."

Caption from photo of Bill Lawrence in a tree looking for moving water:

"Our task for the day was to discover the (or a) head of navigation for the Wolf River. We were able to get this far across a swamp that we accessed from a logging road that followed a high ridge behind me (behind the photographer) and we knew that the channel of the Wolf River was somewhere ahead in the mist. Bill took the high road and looked for some type of "path" for our canoes. I took the low road and slowly drifted around the area with my face hanging out over the water trying to find traces of current. It was like watching a hand on a clock that may or may not be running. We found a path about 30 minutes later."

Caption from photo of two men pulling boats over fallen trees in a swamp:

"This swamp on the Wolf River is about a mile upstream (due south) of the U.S. Hwy. 72 bridge in Benton Co., Mississippi. There are many

RIVER ANGELS

other such beautiful wetlands along the Wolf River, but this was the hardest one to penetrate.

This was also the hardest day of canoeing in my life. Fitz and I were joined by Ray Skinner and Bill Lawrence (blue shirt), who is something of a Yoda or Ben Kenobe figure in the uppermost Wolf and an invaluable guide to us for this section. We pulled our gear-heavy canoe out of the shallow water and over fallen trees almost every 150 feet of river channel. We only made five miles that day. It rained its butt off that night, which was good. Come Hell or high water, I'll take the latter."

Caption from photo of a dead beaver in a trap:

"Trapped, but not collected, by a local landowner. This is a case of pest eradication (small scale) and not hunting. The beaver presents the canoe-owning conservationist with a dilemma. They are part of the native ecosystem, but they seem to despise moving water, so they build dams (and I spelled it "damns" at first) everywhere they can. Landowners may disagree with the beaver's choice of which trees to cut down, or which patch of land needs to be converted into a beaver pond. The beaver population along the Wolf River used to be controlled by its predators, like wolves, black bears, and beaver pelt merchants. They are too big for any kind of raptor, and coyotes don't seem interested. We're too far north for alligators. The Ghost River section of the Wolf River (3 sections downstream of where this was taken) has such vast expanses of still water that the beaver don't seem particularly interested in blocking the channel. Also the streamflow is high enough there that a dam would have to be pretty big to work."

Caption from photo of the Old Highway 72 bridge:

"Near Canaan, Mississippi, 80 miles from the Mississippi River: This ruined bridge over the Wolf River (slated for replacement) is part of Old U.S. 72, which is a dirt road. It's about a half mile upstream (south) of the "new" U.S. 72. This was a welcome sight, as it took 9.5 hours to reach

this point from our put-in at at the provisional head of navigation just 5 miles behind us."

Bridgman's account proved to be an interesting read. It confirmed the reports that Dale Sanders, John Hyde, and others had related: a full descent of the Wolf River would be no picnic of a paddle.

Meanwhile, back in Bikecar land, I had a few shipping details to sort through.

RW: *Hey Paul, questions about dimensions of the bike car. I'm looking at shipping the bike form Eugene to Ontario. What are the approximate dimensions of the bike? How much does it weigh approx? Can the frame be broken down at all?*
PE: *How come ya getting it shipped all the way out there haha.*
RW: *I'm moving to Ontario at the end of this month, at least temporarily.*
PE: *Its 1.7m wide, 2.6m long, 1.4 high and weighs about 100kg. The seats can be removed and laied down.*
RW: *Did you box it or build a crate when you brought it over from England?*
PE: *It got shipped over like it is now, in a container.*
RW: *Is that 100kg without gear?*
PE: *Yes. 100kg without gear. It has about 6 spare tires if I'm right. Make sure you get them too. And there is a spare rim.*
RW: *Can the frame be broken down?*
PE: *Not really. Its all welded together.*
RW: *Is it steel or aluminum?*
PE: *Aluminum. It* (the bike) *can be broken down.*
RW: *Okay, how would break it down? Does the steering apparatus easily detach?*
PE: *Yes, its very easy. Its just a few bolts here and there. I made the bike very simple so parts can be easily replaced from your standard hardware shop. You can take the wheels off, fold the seats down, remove the handle bars and steering rack. It can be broken down to about 0.7m high.*
RW: *The frame is essentially one piece, correct?*
PE: *The frame is all one piece. Here's a video of me building the bike*

(sends two YouTube links). *It might be easy for you to see. You get some idea from them on how it was built.*
RW: *Cool, thanks. Chat later.*

The videos were a huge help in understanding the Bikecar's construction. I forwarded the links to Paul Adkins, a bicycling advocate in Eugene. Paul had caught wind of our storage quest via the International Bicycle Fund and assured me he would look after it. If the bike needed to be broken down and shipped, which is what most of the shipping companies were telling me, then Paul Adkins would be the guy doing the work.

I drafted an email and sent it to Dave Cornthwaite, Paul Everitt, and our new friend, Paul Adkins.

Greetings gentlemen,

I am sending this email to all of you to ensure that we are all on the same page.

Paul E and Dave - Paul Adkins is a bicycling advocate who lives in Eugene, within riding distance of where the bike car is currently stored. I spoke with him over the phone today. He is willing to let us temporarily store the bike free of charge at his place in Eugene. Paul would like to get some information on what our bike car plans are so he can gauge how long he will be storing it. Paul A says he could accommodate the bike until June, but in order to continue storing it he would need to know what our concrete plans are. I mentioned to him that Paul Everitt is planning a trip to the U.S. this summer.

Paul E - are you planning to use the bike this summer before or after your Mississippi trip?

Dave - Any idea when you would need to use the bike?

Paul A - I sent you contact info for Jeff Lozar (the guy currently storing the bike car). I will leave it up to you and him to work out a time and date for you to retrieve the bike. I have sent him your contact info.

I am planning to go to Montana in June. There may be a possibility that I could drive to Eugene to deal with the bike. Ideally, I'd like to store it in Vancouver. Even more ideally would be to store it in Ontario where it would see some use.

Paul A - we certainly appreciate you offering us a space for now and please know that we will remain committed to finding a home for the bike car. If for any reason the bike becomes an inconvenience to you, please don't hesitate to contact one or all of us and we will do our utmost to resolve the issue.

Cheers,

Rod Wellington

Paul Adkins replied:

I will work to get the Bike Car into my yard this coming week.

Dave Cornthwaite replied:

Good morning all,

Rod, thank you for the introductions, and Paul A many thanks for storing the bikecar short-term. I'm looking into options to undertake a journey later this Summer but it's too early to confirm.

Chances are this won't happen by June for me, but will endeavour to find alternative accommodation for the bikecar in that event.

Cheers for now

Dave

With the adventurous among us all on the same page, it was time to contact Jeff Lozar, the gent in Eugene currently storing the bike. Lozar and I had exchanged a few Facebook messages and I got the feeling he was beginning to question whether I'd actually sent him the $100 for the bike and the February storage fee. Perhaps the fact that I'd found somewhere else to house the Bikecar would ease his mind.

Hi Jeff,

Paul Adkins is a bicycling advocate who lives in Eugene, within riding distance of where the bike car is currently stored. I spoke with him over the phone today. He has offered to store the bike at his place in Eugene. Paul A has your contact information. I will leave it up to you and him to work out a time and date for him to retrieve the bike. Please find Paul's contact info below.

Cheers,

Rod Wellington

Six days later, after he'd received the money, Lozar sent me a short message.

Rod,

Received your mail today – thanks

You can have the local guy contact me about picking up the bike.

I was glad to hear the money had arrived, but why hadn't Paul Adkins contacted him in the meantime? Six days and no contact? That seemed strange. After all, they lived in the same town. I decided a nudge was in order.

Hi Paul,

Checking in to see if you've had a chance to pick up the bikecar yet. If not, here's the contact info for Jeff Lozar in Eugene: Office: ***-***-****. Cel: ***-***-****. I've been in contact with Jeff this week. He knows you will be contacting him soon. Dave Cornthwaite and I are looking at options for having the bike shipped to either Ontario or Memphis in March. We will update you with details as they arise.

Cheers,

Rod Wellington

Paul Adkins replied:

Sounds good. I'm chatting with Jeff about a pick-up this weekend.

Paul

"Well, that's a relief," I thought. "Everything seems to be on track."
   In the meantime, ol' Corn was brewing up a new adventure idea that would link into his Expedition1000 project and seamlessly tie all this loveliness together.

DC: *Hello buddy. So, Memphis to Miami is 1001 miles. Fate, surely? Can you think of any cunning ways to get the bikecar to Memphis in time for us wrapping up on the Wolf??*
RW: *Perfect! I was thinking of getting it shipped to Ontario. I will be in touch with the guy that's storing it*—will *be storing it*—*and ask him if it can be broken down and shipped. Then it's just a matter of getting it to Memphis. I may be driving down there (from Ontario), so that could work. I was thinking yesterday that it would be good to have the bikecar in Memphis for school talks and the like. Kids would love that thing! And adults too!*
DC: *Haha, do you think you will drive or fly down? It would be great to*

have for school talks, and I have some fresh air in April/May so perhaps a little 4-wheeled journey straight afterwards could be just the ticket.
RW: *Cheapest option for me is to fly out of Detroit, then use Dale's Toyota while in Memphis and a paddleboard on the Wolf. That way I don't have to bring the truck or my kayak with me. That's an idea I've been mulling over. However…It may be possible to ship the bikecar directly from Eugene to Memphis. Or…I could ship it to Ontario and drive it down to Memphis. I had been wanting to use it around Chatham while I am there. It'll be a good self-promotion tool while I'm there. Either way, I can promise that I* will *get it to Memphis. So, if you want to commit to the bikecar (BikeCar sp?) journey from Memphis to Miami in April/May, I can commit to getting the BikeCar to Memphis.*
DC: *You're the best Rod. I'm shaping ideas over here. Memphis to Miami is tempting but dependent on a few things, opportunities arising etc. Will keep you posted and let you know ASAP. I can't wait to try the bikecar, it looks phenomenal. Will chat soon Rod, keep on keeping on.*

Meanwhile, in Memphis, the SUPAS planning was progressing at an equally ideal pace.

On February 16, 2012 at 12:25AM, Jonathan Brown wrote:

I'll keep this short and sweet…

Our official team is as follows:
Richard Sojourner, Dale Sanders, Luke Short, Adam Hurst, Dave Cornthwaite, Rod Wellington, John Hyde, and Myself. (8)

TEAM MEETING: I'd like to see if we can all gather soon. I was thinking this Saturday or Sunday…. Let me know what you think soon. I'd also like to see if we can get Dave and/or Rod in this via Skype. This will be a rather informative meeting with good convo I'm sure so hopefully everyone can make it.

Rachel with OBS is back from Cambodia and will be carrying the event on from here. Robinson is still rocking it on all the logistics of the event while Rachel will be responsible for the official stuff and media.

Wolf River Conservancy, understandably, will not be able to officially partner with this event but will be offering help and support where needed.

Richard Sojourner will be helping out with getting the Bluff City Canoe Club on board helping out at the stations and elsewhere. On that note… Richard… I just taught Patty McLaughlin today and she said she would like to be involved.

DAVE and ROD:
We can't tell you enough how much we area all stoked on you're participation!!! As official as that just sounded, the bottom line is that we can't wait to kick it with you guys!!!

Please communicate with Rachel about utilizing your social networks to help get the word out and run updates through. They are working on their side to get a blog-styled site up. If you have any recommendations and ideas please shoot them her way.

If you haven't filled out the Aquapac Order Form please do so soon using the link below. John Hyde brought it to my attention that I missed a few things on the form such as a SLR Stormproof camera bag. Anything you want that's not on the form just write that in the "comment" section.

Also, send Robinson your mug shot if haven't done so already.

SIDENOTE:
We hope you've been able to spread the word about SUPAS. Rachel and Robinson are working on press releases and official communication point to ensure we're all on the same page when we share with others the purpose and vision for this event. Also, for now, we ask that you not talk so much about "where" we will be paddling in detail. If you can keep it vague like, "We're paddling from the source to the mouth of the Wolf." There are some

sensitive issues that need to be worked out. We'll explain more when we all hang out.

Okay... not so short and sweet but please respond with any questions concerns or additional communication.

Mahalo!!!

Jonathan

On February 16, 2012 at 12:25AM, Ray Graham wrote:

This is the blog site and youtube page with entries concerning the HWY 72 to Michigan City last weekend. This was an impromptu float, so sorry if we missed anyone. There is another trip to this area planned for March 14th. Please email mail if you are interested.

Ray

www.wolfriverguide.blogspot.com

www.youtube.com/user/wolfriver100

  Ray Graham—a certified river guide with the Wolf River Conservancy—grew up beside the Wolf and knew it well. He'd been involved in the SUPAS planning from the beginning and his intimate knowledge of the river provided the other team members with a wealth of information. Ray regularly captured his Wolf River adventures with video and words. His blog and YouTube channel were filled with astute flora and fauna observations, as well as the changing moods of the river in any given season.

  On February 12, 2012, Ray, along with Jonathan Brown, John Hyde, and Dale Sanders, braved bitter winter conditions when they paddled the Wolf from the Highway 72 bridge to the town of Michigan City, Mississippi, a distance of about five miles. This was the first of several scouting missions undertaken by members of the SUPAS team. John Hyde, in a fitting tribute to early American explorers Lewis and Clark, dubbed the team "The Corps of Discovery."

Even though I hadn't met John Hyde in person, certain clues in his emails easily tipped me to the fact that he was a man who possessed an audacious attitude toward anyone who challenged his knowledge of the upper Wolf River. To his credit, he'd not only obtained his understanding of the river via many hours of paddling its languid waters, but he'd also spent time in the presence of a master—the "Yoda of the Swamp" himself: Bill Lawrence. With Lawrence now gone—he died in 2011—John assumed the role of Swamp Master. He'd even purchased Lawrence's canoe and paddled it with pride. He nicknamed the craft the *U.S.S. Yoda*.

When SUPAS members needed to learn more about the upper Wolf, they sought out John Hyde. His name had been given to them by Keith Kirkland, Director of Outreach at the Wolf River Conservancy. Keith was old friends with Bill Lawrence, Larry Smith, and Gary Bridgman and had paddled with these men many times. One could say that John Hyde was more of an *acquaintance* of Keith's, not a *friend*. John, like Bill Lawrence, led a quiet life. He kept to himself and enjoyed the solitude of the swamp. Keith was certainly familiar with John Hyde's eccentric idiosyncrasies (one could say that John is *not* a team player), but, as a paddler who was passionate about the Wolf River, Keith also had ample respect for John. Keith certainly had this respect in mind when he forwarded John's contact information to the members of SUPAS.

John Hyde, as I learned later, had a deep connection with the Wolf River. He'd swam in it as a child and paddled most of its length. (As of 2006—six years before SUPAS—he'd paddled every mile of the river from Blackjack Road to its confluence with the Mississippi, a distance of 96 miles.) John was plenty proud of that feat, but it was the six-mile section upstream of Blackjack Road—a section he'd never seen—that now captured his interest. To the best of his knowledge, no one had paddled this short section. John's mentor, Bill Lawrence, who had experienced that section on foot with Bill FitzGerald and

RIVER ANGELS

Gary Bridgman in 1998, cautioned his young river apprentice when the two exchanged communication in order to determine the possibility of paddling from Baker's Pond to Blackjack Road. "I wouldn't try it if I was you, John," said Lawrence, "you might get shot! There's fences across the river and people *don't* want you there!" When Bill passed away in 2011, John knew he had to paddle it, and he knew he had to do it in Yoda's canoe. When the SUPAS team came calling in early 2012, John's chance to finish the final section of the Wolf had materialized.

"It's gonna be a real motherfucker of a paddle, that's fer shurr," thought John. "But when I float under that bridge at Blackjack, I will become the first person to paddle the *whole* Wolf, all the way from its source to the cobblestones at the foot of Union Street in Memphis. Ain't nobody gonna be able to say that but me!"

Dale Sanders was also deeply connected to the Wolf. As a river guide with the Wolf River Conservancy, he'd spent hundreds of hours paddling and exploring its length. He prided himself as someone who knew the river very well. One could say he was somewhat of an expert when it came to the Wolf. But, like John Hyde, there was one section of the river that Dale didn't know, one section he'd never paddled or even laid eyes on—the remote section upstream of the Highway 72 bridge.

The seemingly impenetrable swamp upstream of Highway 72 was daunting to say the least. The Wolf River Conservancy viewed this 20-mile section as dangerous and off limits to paddlers. According to the Mississippi state law, that portion of the Wolf was unnavigable. It passed through a stretch of snake-infested swamps and was bordered on each side by private property. The owners of said property didn't like trespassers and they aimed (quite literally) to keep them out. This *was* the South, after all—a place of whiskey stills, shotguns, and intolerant, trigger-happy landowners. Bill Lawrence knew the risks and acted accordingly—he stayed away. Paddlers like Dale Sanders and John Hyde, however, don't listen well to words of caution. They have other ideas in mind—*their* ideas.

Dale knew full well that John Hyde desired to become the first person to paddle the Wolf end to end, and John was quite keen to remind Dale of that fact as often as possible. Dale, however, had other plans. He wasn't content to sit back and watch the title go to John Hyde. Dale was a well-disciplined Navy man who'd won prestigious awards at sporting events and had held world records. At age 76, he'd had a long life of travel and tribulations. He wasn't about to give up anytime soon.

If there's anything Dale Sanders truly loves, it's *competition*. He'll be the first to tell you that. The man loves (and *lives*) to win. Dale knew full well that no one had ever paddled the entire length of the Wolf River. When the SUPAS team came calling in late 2011, Dale knew that his chance to add another prestigious title to his cache had finally materialized. Now, with John Hyde's addition to the SUPAS team in early 2012, Dale's easy ride to victory just got complicated. Like it or not, Dale had a competition on his hands and he had no intention of losing.

Their feud was evident to the other SUPAS members and it was starting to leak into email correspondence being shared with everyone involved in the project. Their face to face meetings regularly became less-than-lighthearted debates about the Wolf River and its environs. John both supported and opposed several Wolf River Conservancy initiatives. Dale, of course, took the opposing viewpoint. Thankfully, they steered away from politics and religion—they were polar opposites in those arenas—but found ways to craftily goad each other on nearly any subject under the sun.

On February 12, 2012 (the day John Hyde, Dale Sanders, Jonathan Brown, and Ray Graham paddled downstream from the Highway 72 bridge to the town of Michigan City, Mississippi—the first of several SUPAS scouting missions), Ray observed Dale and John bickering about trivial minutiae and made mention of it in his corresponding blog post.

"We stopped for lunch and Jonathan found a large sheet of ice in a nearby stream. He held it up to the sun, and John and Dale took pictures of the sun reflecting off the ice. The pics were unique, and I hope to get a copy. It was funny to watch John and Dale give contradictory directions for the

fire and ice poses. Finally, Jonathan had enough and smashed the ice sheet over his head."

To his credit, Ray managed to relate the riverside exchange between Dale and John in an amusing way. From Ray's perspective, the absurdity of the scene seemed "funny." But lurking beneath the unintentional humour was an unspoken darkness that Ray illustrated in his "fire and ice" observation. What he and Jonathan were witness to that day was a "contradictory" dance between two opposing forces who, in many respects, were very similar. Their childish bickering about unimportant camera angles and how their impromptu "model" should pose for them confirmed the presence of a simmering feud, an undeniable hostility that would only lead to bad blood between river brothers. The battle, it seemed, was just beginning.

Below, in its entirety, is Ray Graham's blog post from February 12, 2012.

2/12/12

On this clear, cold day, a group of us decided to drive to Mississippi in order to attempt the river section from HWY 72 to Michigan City. This area is close to the headwaters of the Wolf, and the upper most section that I have paddled. Our group was small but experienced and include Dale, the venerable river rat; John Hyde, the most experienced paddler on the upper extreme of the Wolf; Jonathan Brown, a pioneer of Wolf River stand up paddle boarding and myself, former WRC intern and volunteer river guide. The sky was a spotless blue, which appeared even deeper through my sun glasses, with temps in the 20s and slowly rising into the mid 30s. The water was relatively clear and deep enough to float. We parked the truck on a dirt drive on the south bank of the Wolf, east of HWY 72. The drive is easily accessible off the highway. I was relieved to see that we could park away from the highway traffic and that we were also close enough to the river to avoid a long portage to the bank.

At HWY 72, the river is close to 50 feet wide. However, this is said to be the upper most point on the Wolf that is legally navigable. Oddly enough, state law says that a flowing body of water with a volume of 1000 cubic feet per second is considered navigable. This would imply that the Wolf is legally navigable above the HWY 72 bridge. Based on conversations with the locals, anyone paddling upstream of HWY 72 is considered to be guilty of trespassing. I'm paraphrasing the law and emails I have read. To the best of my knowledge these statements are true.

The temps were well below freezing when we launched. The banks were lined with hoar frost, ice crystals pushing up through the river mud. Surface ice covered many of the sloughs and creeks that entered the Wolf. I sliced through the ice in my otter kayak just to hear the crackling sound of the ice breaking against the bow.

Our group moved down the main channel easily. The water depth was not a problem, but we did have to steer around partially submerged logs. Since they had floated this section before, Dale and John took the point position. I lagged behind to take video. As we followed the main channel, I saw a pair of great blue herons off to the right. We were also following a flock of about twenty ducks downstream. They would burst into the air as we approached. We saw them about every 30 minutes as we paddled with the current.

The right bank of the main river channel was marked with signs designating the land as a natural area. We could also see that the land to the right dropped away and actually appeared to be lower that the riverbank. This was a hint that at some point the water would spill over the bank and leave the old river course. A little while later we could hear water spilling through the brush on the right bank and then the main channel terminated in a grassy swamp. To the right, we saw a sign that read "Smith Corner". This was the domain of Larry Smith, a WRC founder. He owns some acreage on the right bank and even put some blue canoe trail signs. We had to backtrack approximately 200 yards to find a turn off on the right bank. The water led us through the trees and before rejoining the main channel. We followed the old river channel at times, but there were at least three turns that we had to make where the river *left the channel. We also had to portage over some downed trees, but it wasn't as bad as I had expected it to be. I had prepared myself for a portage every hundred yards. Luckily, we had a great deal of clear paddling. At one point, we took a right instead of staying with the main channel. The tributary we were in became almost to shallow to float. I hiked north away from the group to look for another channel, and saw flowing wet lands that continued to a high red dirt bluff almost a hundred feet high. When I returned to the group, we decided to continue forward, and we were rewarded by finding the main channel again. It was even marked with the blue canoe signs.*

Besides ducks and great blue herons, turkey vultures circled overhead most of the trip. Other birdlife included a variety of woodpeckers including a redhead, hairy and a downy; cardinals and sparrows. I saw a raccoon climb down a tree on the right bank and swim across the river directly in front of us. I marveled at the animal's ability to swim in sub freezing temps. Wildlife signs included deer tracks on the bank and of course beaver dams. We also saw signs of human use. We passed under an old iron footbridge. The group also saw remnants of a wooden bridge, deer stands and even a rope swing. Dale was going to try the rope swing but declined since he didn't bring swim trunks. Paddlers should be aware that much of the lands to either side are private tracts. Care should be taken not to trespass.

We stopped for lunch and Jonathan found a large sheet of ice in a nearby stream. He held it up to the sun, and John and Dale took pictures of the sun reflecting off the ice. The pics were unique, and I hope to get a copy. It was funny to watch John and Dale give contradictory directions for the fire and ice poses. Finally, Jonathan had enough and smashed the ice sheet over his head.

We floated of a number of rapids, some caused by cypress trees, some from beaver dams. We had fun with the largest one as we attempted to paddle back up it. Each of us tried to fight the current, but we failed. On my second attempt, I was almost over the lip of the rapid when water began to spill over my bow and into the cockpit. Fearing being swamped in sub freezing weather, I decided to give up.

Towards the end of the section, John directed us to follow a stream course on the left side of the river. It led to a large beaver dam, the largest I have seen on the Wolf. We marveled at the engineering. The curvature of the dam allowed it to hold back a great deal of water. It was the same concept as a Roman arch except it was lying on its side. The water on the other side of the dam was about four feet higher than the streamside. John remarked that, "Beavers are pretty smart for being so low to the ground." How true.

A short while later, we reached an old, wooden train trestle and HWY 7 near Michigan City. I was relieved to find that we could take a right up a canal and easily land next to Dale's waiting van.

It was a fun trip and easier than I had expected. However, there are dangers to paddling in such cold weather. If one of us had capsized, he could have faced a serious threat from hypothermia. In addition to being well dressed, we were all experienced paddlers. I was wearing waders, two pairs of wool socks, two pairs of nylon pants, four layers of nylon and wool shirts, a water proof jacket, water proof gloves and a warm hat. A combination of the sun, clothing and activity made it feel like a warm spring day despite the chilly temps. Because of the cold, there was also the tendency to become dehydrated. One should not to forget to drink water even in the cold.

On February 17, 2012 at 12:47AM, Jonathan Brown wrote:

Hey guys, I haven't heard back from all of you but if we want Dave in on our meeting then it will need to be around 10 or 11am on Saturday… that's tomorrow. Richard, Dale, and John said their cool with any time. Luke will be out of town. Have yet to hear back from Adam, Rod, and Ray. Also… Rachel and Robinson, let me know if you'll be able to make it.

The goal is to get everyone together so if this time doesn't work please let me know Friday aka today. :)

Jonathan

On February 17, 2012 at 8:59PM, Jonathan Brown wrote:

So it's decided! 11am tomorrow morning it is! I'm still working on where it will be. Bear with me like a grizzly and I'll let you know where it is soon. I'm thinking Republic but need to see if we can have a conference room to set up Skype.

Rod and Dave will be able to Skype for sure so that will be great!!!

ROD AND DAVE: Once i figure out where I'm thinking you and I need to do a test call 20 min. before to make sure there's not issues. WOuld you two be cool with that?

JB

On February 17, 2012, Dave Cornthwaite tweeted:

Global skyping with folks in Memphis and Vancouver cooking up a world first SUP paddle for mid-April @rod_wellington @SUPAS_WolfRiver

On February 21, 2012 at 12:16PM, Dave Cornthwaite wrote:

Me again! Might be worth dropping John Ruskey a line to see if we can borrow 3/4 of his Stand Up Paddleboards. They're just the type we're after (see pic from my trip)

Jonathan Brown had been hard at work trying to score sponsorship assistance with a stand-up paddleboard company. Most of his email inquiries went unanswered. Some companies, like C4 Waterman, had offered sizeable discounts on purchased boards, but no one had stepped up and offered boards outright at no cost. Of course, it wasn't in our favour that we only intended to use the boards for one week. Based on my experience of dealing with watercraft manufacturers, they generally only assist those undertaking long-distance journeys, like a source to sea trip down the Mississippi River. So, the chances of receiving four sponsored SUPs from a North American manufacturer for a week-long river descent didn't look too promising.

During Dave Cornthwaite's SUP descent of the Mississippi River in 2011 (sponsored in part by Lakeshore Paddleboard Company), many people living along the river came forward to offer assistance—a hot shower, a dry bed, a home-cooked meal, shouts of encouragement from the riverbank. For reasons unknown, "People on the river are happy to give."

Selfless acts of generosity are not limited to one journey on one river, however. Kindness of the like that Corn experienced happens worldwide. Anyone who has travelled by self-propelled means (or otherwise) has had at least one random act of kindness bestowed upon them. Books could be filled with the myriad of wonderful stories told by grateful travellers.

Long-distance hikers on the Appalachian Trail (a 2200-mile-long trail in the eastern U.S.) often talk of their love for "trail angels," kind-hearted folks who live along the trail and offer support to those making the long trek. When asked what inspired their decision to give of themselves altruistically, one trail angel responded, "The one goal that everybody should have in their life is that they are more of a giver than a taker." Another said, "I'm helping them, but they're helping me at the same time." Unintentional reciprocity builds trust and grows love. Strangers become friends. Friends become networks, and networks play huge roles in the success of projects. Opening oneself to new opportunities often leads to amazing things. A simple "Yes" can open many doors, especially when it comes to building solid relationships. As ol' Corn would put it, "Strangers are just friends waiting to happen."

Norm Miller lives in Livingston, Montana, a stone's throw from the Yellowstone River. He grew up in Michigan, moved west to work at Yellowstone National Park at the age of 21, and has lived near the Rocky Mountains ever since. He wouldn't have it any other way.

I interviewed Norm at his home in Livingston in December 2014. Sporting a pine-coloured ball cap over his closely cropped hair, and

RIVER ANGELS

clothed in outdoorsy attire straight out of an REI store, Norm looks like the kind of guy you'd meet atop a blustery mountain peak, or careening down a fast-flowing river, or even hiking the Appalachian Trail. He's an adventurous soul more at home in nature than in the asphalted terrain of a city.

Norm still speaks with the lazy Midwestern twang of his birthplace. His tone is flatly serious and loaded with details—dates, distances, dimensions. Our conversation leans mostly toward American history and canoeing, two subjects that cater to his obsession with details, logistics, and the great outdoors. If something significant happened near an American river, and it had something to do with paddling, Norm probably knows about it. It's not a stretch to say that Lewis and Clark's monumental journey up the Missouri River in 1804, and their subsequent return to St. Louis two years later, figures strong in any conversation with Norm Miller.

In spring of 2004, on the bicentennial of Lewis and Clark's departure from St. Louis, Norm began his own solo paddling journey up the Missouri, retracing the route of the Corps of Discovery. His expedition spanned 199 days, including a five-week backpacking trek through the Rocky Mountains before continuing to the Pacific Ocean on the Clearwater, Snake and Columbia rivers.

When you paddle the longest river in North America *up*stream *and* end to end, you become somewhat of an expert about that river. Intentionally—or perhaps *un*intentionally—you make a name for yourself. Paddlers search you out and pick your brain. You become known as the "go-to" guy when it comes to planning a paddling descent of the Missouri River. People want to know what *you* know. They want facts, figures, and the location of your favourite campsites. They want *details*, and Norm is the perfect guy to give it to them.

You see, Norm is a "river angel"—a term he allegedly coined, an adaption, he says, of the venerable "trail angel" term associated with the Appalachian Trail.

One of the people who sought out Norm prior to his journey up the Missouri was a dugout canoe carver named Churchill Clark. Churchill, as it turns out, is a direct descendant of William Clark, he of Lewis and Clark fame. Churchill was readying himself for an ascent of the Missouri River in one of his dugouts and approached Norm for logistical assistance. Norm remembers well his first exchange with Churchill.

"He calls me basically a week before I'm leaving on my trip in '04 and says, 'Are ya interested in having somebody kayak with ya?' and I'm like, 'No, I wanna do this solo.'"

Norm, however, didn't completely rebuff Churchill's query. Knowing that Churchill would be interested in a group of history enthusiasts who were re-enacting the Lewis and Clark expedition that same year, Norm passed along their contact information.

"I said, 'There's this group out of St. Charles (Missouri) that's recreating the whole two-and-a-half years. Here's the phone number. Get going!' And he did. He spent the next two-and-a-half years with them."

In spring of 2006, Churchill returned to Livingston, Montana with the re-enactment group. Accompanying him were two experienced river guides who would go on to greatly widen Norm's burgeoning "river angels" network.

"They arrived here (in Livingston) on horse from Orofino, Idaho. Churchill had been on a horse for about a month. And then him (Churchill), Mike Clark (no relation), John Ruskey, and about five others built two 28-foot dugout canoes about four blocks from this house, and paddled them to St. Louis."

Norm joined them for a couple days in his kayak on the Yellowstone River and, later, joined them again for the last seven-day stretch of the Missouri River to the Gateway Arch in St. Louis. The four bonded well on the river and became fast friends.

Years later—in 2011—when Norm created a Facebook group page called Missouri River Paddlers (which now contains perhaps the world's most extensive collection of Missouri River paddling information), Mike Clark and John Ruskey were among the first people to join. Ruskey, who lives in Clarksdale, Mississippi, went on to create a complimentary Facebook group page entitled Lower Mississippi River Paddlers, partnering with fellow paddler, John Sullivan, who

RIVER ANGELS

admins the very popular Mississippi River Paddlers page—all great resource pages for anyone planning a paddling journey down (or up) the Mississippi River.

Paddlers interested in doing the Missouri and Mississippi rivers regularly contact Norm, Ruskey, Sullivan, and Mike Clark for logistical assistance. It's not uncommon for any of these four gents to join paddlers on the water for a few days, or offer paddlers a warm, dry place to lay their weary heads for a night or two. Collectively, these four river vets have paddled tens of thousands of miles, so they know very well that river travel can be both joyful and tiresome. There comes a time when even the toughest paddlers need respite. And when they do, they know, if they've done the research, that a flourishing network of "river angels" is waiting in the wings to help recharge their batteries. It's not uncommon for river angels—who learn about paddlers and their journeys through close-knit networks on social media—to extend offers of help long before paddlers ask for assistance. For many paddlers, Norm Miller, John Ruskey, John Sullivan, and Mike Clark are their introductory links to this network.

Such was the case with Dave Cornthwaite. Prior to his SUP descent of the Mississippi River in 2011, ol' Corn sent out emails to dozens of canoe and kayak clubs up and down the Mississippi. The emails contained info about the upcoming expedition as well as a polite request for possible assistance along the way. The response was overwhelming. People replied not only with offers of accommodation and logistical help, but many of them also pointed Dave in the direction of "Driftwood Johnnie," a storied river rat based in Clarksdale, Mississippi who also goes by the name of John Ruskey. Ruskey, in turn, suggested Dave get in touch with "Big Muddy" Mike Clark, owner of Big Muddy Adventures, a river guide company based in St. Louis, Missouri.

During his stop in St. Louis, Dave and his paddling partner, fellow Brit, Tom Evans, stayed at Mike's famous Kanu House, located near the banks of the Mississippi, a few miles upstream from the Gateway Arch. Over the years, the Kanu House has hosted a plethora of paddlers from over all the world. Stories of Mike's amazing generosity could fill volumes. His self-assigned role as "gatekeeper" of the lower Mississippi River (based on the fact that he's located at the Missouri-Mississippi river confluence) has helped him meet many paddlers travelling down the Mighty Mo and the Mighty Miss.

Ruskey, who owns Quapaw Canoe Company (a river guide company in Clarksdale, Mississippi), hosted Corn and his brother Andy when they arrived in John's neck of the woods in August 2011. In this excerpt from his Wild Miles website (www.wildmiles.org), Ruskey shares his thoughts on the past, present, and future of the Quapaw and its many side projects, including his partnering with Mike Clark's Big Muddy Adventures.

Quapaw Canoe Company provides guided canoeing & kayaking on the Middle and Lower Mississippi River, from St. Louis to the Gulf of Mexico. We are based in Clarksdale, Mississippi, with an outpost base in Memphis, Natchez, Vicksburg and Helena, Arkansas. Big Muddy Adventures is based in St. Louis near the confluence of the Missouri. My name is John Ruskey. I am the founder & owner Quapaw Canoe Company (est. 1998). I first started paddling the Mississippi River with a 5 month raft trip in 1982-83. Mike Clark founded Big Muddy Adventures in 2001 with a 3-month educational expedition down the Mississippi. We guide thousands of visitors every year on the Mississippi River, mostly on overnights and multi-day trips through remote sections of water with little or no industry and agriculture. We employ dozens of guides, shuttle drivers and support staff. We engage and employ local youth through long term-apprenticeships on canoe building, canoe technique & rescue, wilderness survival, and the business of nature-tourism. We help maintain the health of the river system with cleanups and educational campaigns.

We at Quapaw Canoe Company and Big Muddy Adventures are very concerned about the detrimental effects of any future industry along the Middle (St. Louis to Cairo) and Lower Mississippi Rivers (Cairo to the Gulf). Future power plants, steel plants and refineries are of particular concern, but also grain elevators and hydrokinetic turbines. We are concerned about the remaining wild places along the Middle & Lower

RIVER ANGELS

Mississippi River that might be compromised with any new developments industrial or agricultural.

Our clients come from all over America and all over the world to see the Mighty Mississippi River. We call it "America's Forgotten Wilderness" because the river channels are so big & open, the islands are wild, and there are extensive unbroken forests throughout. There is abundant wildlife. It is one of the last remaining habitats for Black Bear in the center of the country. The Mississippi is the most important flyway in North America. There are hundreds of fish & amphibians. It is our strongest & most vital inland fishery. Most importantly to us and our clients when you are paddling the river it truly feels like a wilderness. It's big and open. Your imagination is intrigued by the scale of the landscape, similar to what you might feel in the Rocky Mountains or the wilds of Alaska. At night, the stars are visible much brighter than anywhere else in the Middle of America.

There is towboat activity, of course, but towboats come and go. Any permanent industrial installations within the Wild Miles would change the feeling of wilderness for paddlers & campers with visual and audible pollution. In today's busy world it is difficult to find wild places where you can leave behind the trappings of civilization and reconnect with those aspects of nature & survival that is so important to the American pysche as defined by Thomas Jefferson and championed by Theodore Roosevelt.

The floodplain of the Lower Mississippi is one of those places that the feeling of wilderness has been preserved —why?

By the power of this dynamic river. Regular high water events have kept most development away. The river fluctuates 40-65 vertical feet in any given year, high water covers most islands and floods most forests & batture areas (the river side of the levee). While the seasonal flooding prohibits industry & agriculture it actually enhances its value as a wilderness - especially for those who can reach its hidden places by water. It is a paddler's paradise. It is one of America's greatest paddling challenges. We are certain that it will become a classic destination for canoeists & kayakers alongside popular destinations such as the Boundary Waters, the Adirondacks, and the Everglades.

We enjoy the river as a "forgotten wilderness" and approach it as such with multi-day expeditions by canoe or kayak and primitive camping on remote islands, towheads and sandbars found flung along the banks of the approximately 1200 miles of free-flowing river downstream of St. Louis. Big Muddy Adventures of St. Louis provides the same along the last 340 miles of Missouri River below Kansas City.

After spending a few days visiting with John Ruskey during his SUP trip down the Mississippi River, Dave Cornthwaite (along with his brother, Andy Cornthwaite) continued downstream with John and several of his Mighty Quapaws (river guide apprentices). Their mission: trash cleanup along the riverbanks. Videos of the cleanup showed the Quapaws standing atop 12-foot, multi-coloured, plastic SUPs. John had 10 of these boards in his Quapaw Canoe Company fleet. Manufactured by YOLO (You Only Live Once), a Florida-based company, the boards were solid, stable, and perfect for a week-long jaunt down the Wolf River. Thanks to Corn's networking magic, John generously offered SUPAS team members the use of four paddleboards from his Quapaw fleet. We were more than thrilled to accept. It seemed that our friendly "river angels" had smiled on us yet again.

On February 23, 2012 at 8:06PM, Dale Sanders wrote:

Hi John (Ruskey), *we met briefly several years ago. Keith Kirkland introduced us at one of the Memphis-in-May races. We have several mutual friends, among them Dave Cornthwaite.*

Believe Dave mentioned, I would be contacting you regarding picking up three of your SUP boards for use in the up-comming Wolf River source to Mississippi River paddle. We plan on starting our, down river adventure Saturday, April 7th, day before Easter Sunday. Best on this end, would be pick the boards up anytime Monday 2, April? Or since Rod Wellington will be arriving on 3rd, that might also work. I could return the boards on Monday or Tuesday April 16 or 17. What are your thoughts here?

Rod will be conducting a slide show presentation to Bluff City Canoe Club members, evening of April 4th. Of course you and family etc. are invited. Will send you a special invitation under separate cover.

Believe Dave will be arriving on Friday, April 6th. Will pack and get ready for next day launch. He and Rod will be staying at my place. If your time and schedule permits, please feel fee to drop be anytime you like. Also, Dave will be conducting WRC sponsored presentations Monday and possibly Tuesday after the paddle. I will ask Keith K to keep you informed of dates and times.

You are certainly welcome to join us on the paddle. Wouldn't it be great to have Dave, Rod and you on-board for same event?

Thank you so very much John for helping support our paddle.

Very Respectfully, F. Dale Sanders

On February 24, 2012 at 12:55AM, Jonathan Brown wrote (to John Ruskey):

I wanted to drop a line and say thank you for letting us use your SUP boards. This truly puts the final touches on our event. Swimming wasn't an option.

I've heard a lot about you… good stuff. I would be stoked if you considered joining us on this paddle. I agree with Dale. It would be an epic journey with you Dave, Rod, Dale, and the rest of the guys. I'm sure we would have no problem putting you up for the time you're here. No worries if you can't make it though. Either way I look forward to eventually meeting you to thank you in person.

I must also say that Operation Broken Silence, the organization putting the event together, is very thankful as well. They wouldn't have much of an event without some stand up paddle boards… after all the event is titled, Stand Up Against Slavery (SUPAS). If you would like to now more about the purpose and goal of our event feel free to ask and check out obsilence.org.

Jonathan Brown

On February 24, 2012 at 1:32AM, Jonathan Brown wrote:

Hey guys!

There's been a lot of action happening so it's time for an update.

TEAM CHANGES
I am sad to say that Adam Hurst will not be able to paddle with us anymore. He has school and work obligations he's unable to put on hold. Plus he has a girlfriend and his priorities are right on. ;) The plus side is that he wants to offer support while we're on the water by bringing or taking gear if his schedule permits. He may even drop in on us for a leg or two of the journey.

Team Summary: Dale, Richard, John Henry, Rod, Luke, Dave, and Jonathan (7 Total)…. It will be 8 IF John Ruskey wants in :)

GENEROUS FRIENDS
As we began planning this event we would never had expected Dave an Rod would be joining us. Through their amazing hearts and the most generous help of Wolf River Conservancy we have to amazing men with us! Thank you guys and Wolf River Conservancy (Keith Cole and Ken Kimble)

With that said, since then Dave has hooked us up with dry bags from Aquapac (fill out the order form if you haven't yet), Rod is trying to get a hammock company to sponsor hammocks, and through Dave's friendship with John Ruskey, John has generously allowed us to use his SUP boards for the event!!!

RIVER ANGELS

EVENT INFO
While the paddle team has been focusing on the paddle, OBS has been working on the event side of things. I just had a meeting with Robinson and I must say, I am excited to see what will come of this! Rachel and Robinson have got some great plans in store.

As the team gets closer to Memphis they will be in for some great fun, engaging events, and cool competitions. OBS decided to focus more on the events surrounding the paddle itself and less on a huge closing ceremony. If all goes well, the team will paddle into the boat ramp at Harbor Town April 14th, with a group of college age participants and local media, friends, and family waiting to cheer us in!

Once OBS solidifies the various events connected to the paddle we will send you the details.

WEBSITE & SOCIAL MEDIA
SUPAS now has an official TWITTER and FACEBOOK page! OBS is currently working on the SUPAS website and will be funneling all the photo and video uploads through the OBS Youtube and Flickr sites. They hope to have the site up and running next week! Dave and Rod, I'll connect with you to follow-up on getting the SUPAS Event present on your networks and sites.

Check out the face of SUPAS Facebook and Twitter below:
Facebook: http://www.facebook.com/pages/Stand-Up-Against-Slavery/242282799194178
Twitter: https://twitter.com/#!/obs_SUPAS

Alright, it's late…. If I missed anything please let everyone know.

Mahalo for everything everyone is doing!

Jonathan

On February 24, 2012 at 2:22AM, Jonathan Brown wrote:

Dave and Rod: After I finished putting up the SUPAS site and social network stuff OBS decided they wanted the site to run specifically through their site as well as Youtube and Flickr. Fully understandable but a bit of a slow down in the process. It works out good in the end because now they have a real graphic designer working on it. Robinson has them working on it and hopefully will have it finished by next week. They're keeping the logo, which is good because that took me forever!!!

All that to say, once I can get the material from them, I'll send it to you two so you can begin placing it on your stuff. If you guys have any tips and tricks and ideas for us concerning the social media stuff and ways to get peeps in our network feel free to share.

Also please throw me your thoughts and or ideas about trumping up local media. I'm sure Rachel and Robo are on it but it never hurts to have extra tips and tricks, thoughts and ideas.

I'm so stoked to see how this thing is shaping up.

Rod: You are so right! I am fully relieved the boards worked out!!! Now I can worry about other stuff with this thing. Hey, have you heard anything from the hammock company?

Dave: Bro, you are amazing! Thanks again for the connection with John! Big stress diverted for sure. How soon do you need the order form for Aquapac?

Dale: Thank you so much fro everything you're doing!!!

OK…. I've been sitting in my truck in a parking lot using Tropical Smoothies internet for hours now. My back hurts, I work early, and I need to sleep.

Peace out you three!

On February 24, 2012 2:46AM, Dave Cornthwaite wrote:

Hey JB,

All good. Aquapac order form middle of next week. I'm cranked until then so can't do anything...

Recommend taking Ruskey off all the group emails unless he wants to be a part of the journey itself. He's a busy guy and will have had 10 or so emails bounce into his inbox with roughly the same content, which he hasn't signed up for, yet! Don't want to annoy the guy...

Ok, offski to write more and pretend I have a real job...

DC

On February 24, 2012 at 8:55AM, John Ruskey wrote:

Good Morning Dale, Dave et al:

Dale, of course I remember you — who would ever forget a face as river-grizzled as mine?!!!!

Great adventure you guys have planned down the Wolf — Bravo! I would do anything for my good friends Keith Kirkland and Dave Cornthwaite! I am happy that we at Quapaw Canoe Company could participate in some fashion. Would love to join y'all but we are already booked on the BIG RIVER those first two weeks of April.

Details — let me make sure I understand:

Paddlers: there will be 7-8 paddlers who need boards + paddles?
Dates of Expedition: April 7 – 14
Pick Up: Mon April 2nd (Clarksdale/Helena)
Drop Off: Mon April 16th (Clarksdale/Helena)

(NOTE: we have 10 ABS Yolo YAKs on hand + paddles. You are welcome to as many as you need. You will need several vehicles for transport, or a large trailer.)

Is this a fundraiser for SUPAS? Wolf River Conservancy? Both? If possible, could I get a formal letter of request from the organizations? You can email or mail to the below. I will turn around and see if YOLO wants to participate in this also since all of our board come through them.

By the way — the Yolo YAK — They will be perfect for this, I'm sure: shallow draft, no fin, ABS plastic, pretty much unbreakable...

However, expeditions are always tough on gear. I might ask YOLO if they would cover Quapaw on any damages, loss, etc. I will be sending you almost our entire SUP livery of YAKs for this.

John Ruskey
Quapaw Canoe Company
291 Sunflower Avenue
Clarksdale, MS 38614
www.island63.com

On February 24, 2012 at 11:55AM, Dale Sanders wrote:

Thanks John. Your comments are making me get a big head. The "details" below are correct, except we only need three or four boards. Will know exact number in couple days. To cut down on number of Emails, I will coordinate board pick up and return. My boat trailer will work just fine. Has padded bars for protection against damage.

Jonathan, request Operation Broken Silence (OBS) send Quapaw the support would be appreciated letter. (Please forward this Email on as appropriate). This is not a fund raiser for SUPAS, for we are just bunch of volunteers getting together on behalf of OBS. If funds are raised, all will go to OBS.

I will be paddling my solo Old Town "Pack". My Auto/ Home Owners insurance will cover damages/ losses while transporting boards two and from your place and while here in my home. Once I turn the boards over to the three users, and in the event YOLO will not pick up these possible cost, request Jonathan/ Robinson get the board users and OBS damage/ loss responsibilities clarified.

Understanding your busy schedule, will make every effort, on my end, to limit the number of Emails, to essentials only. Have info'ed Dave & Rod just to keep them informed.

Hey guys, if I missed anything just let me know, I'l fix it ASAP.

John Ruskey replied:

Gotcha Dale! OARSOME. Will await further developments. We'll have the boards + paddles ready for y'all in April.

A pleasure Jonathan — best of wishes on this project, and as we always say around here: "May the River be with Ya!"

John Ruskey
Quapaw Canoe Company

On February 27, 2012 at 12:46AM, Jonathan Brown wrote:

Rod,

I'll be sending out a bit of an update later, but I wanted to let you know that myself and some of the guys surveyed the land at the source and it will not be a ride in the park for sure. I hope you're down for some fairly harsh conditions. :)

It will be so epic to have you, I and Dave on SUP boards for this thing. People are going to flip out and we will def be challenged!

Thanks again bro! Can't tell you enough how much I appreciate you coming down for this! I hope you get everything you need and more in return!

Jonathan

On February 26, 2012, six members of SUPAS drove from Memphis to northern Mississippi to visit Baker's Pond and the headwaters of the Wolf River. Their scouting mission had one simple goal: to view the alleged source of the Wolf firsthand and make some accurate predictions of where to launch their paddling expedition.

Coupling information he'd gleaned from satellite images and conversations with cooperative locals, John Hyde suspected that, contrary to what the Wolf River Conservancy believed, Baker's Pond was *not* the source of the river. John countered with a seemingly provable notion that nearby Goose Nest Lake was truly the utmost source. Armed with an altimeter app on his iPhone and a budding disdain for the WRC, John needed to not only prove his theory to himself, but to his SUPAS teammates as well.

Barely lurking beneath John's rough hide was his budding disdain for his teammate Dale Sanders. Dale was well aware that John needed only to paddle from the headwaters area to Blackjack Road (a distance of six miles) to claim the prestigious title of "first person to paddle the entire Wolf River." John knew that Dale had a longer distance to cover in order to win the prize (about 20 miles). Both men knew that they weren't likely to undertake their individual challenges side by side. They would choose opposing days to paddle this final section of the Wolf, one-upping each other if possible. They would keep the dates to themselves, sharing their plans with the team only at the last minute. They innately knew that doing so could effectively create a wider rift between teammates, possibly spitting the group into two opposing camps. Sadly, neither man seemed deterred by this probable outcome. The day would come, sooner than later, when one man would raise his fist in triumph. The other, in turn, would utter a few choice profanities under his breath, congratulate the victor via email, and move onward with a new mission. Today, however, neither man would paddle downstream. Today, they were there only to gather typographical details while the area was still foliage-free. Their intentions may have remained silent while tromping across the mucky swamp trying to determine the river's true source, but deep down they knew one thing was for certain: the race for paddling supremacy was on.

Below is Ray Graham's account of the SUPAS scouting trip to Baker's Pond on February 26, 2012. Team members in attendance for this scouting trip were Dale Sanders, John Hyde, Richard Sojourner, Jonathan Brown, Ray Graham, and Mike Watson.

2/26/12

Baker's Pond

A group of us drove up to Baker's Pond to explore the source of the Wolf River. John brought his single seat canoe in hopes of being the first person to put a boat into Baker's Pond. We carried the canoe the half mile through the pine forest and over the red loess ridge to the shallow, clear water pond studded with cypress stumps. We climbed the steps up the bluff and followed the trail to the pond's outlet. John put in here and toured the pond. On the return, he found a log protruding from the shore that could provide an easier launching point. Each of us took turns paddling the pond. The log didn't quiet reach the water, but it was easy to slide the canoe through the slick mud to the launching point. The pond was only three feet deep at the most. During my turn, I noticed a fish box in the middle of the pond. Clearly, we weren't the first to float the pond. I didn't see any fish, but apparently someone was attempting to breed them here.

Next we drove to National Forest land near Goose Nest Lake. The lake is private, and we avoided it. However, the road leading past it is public and allows entrance to the forest. As we walked down the road, we noticed several deer stands, a game camera and a steel trap. Anyone entering here should be aware that a large amount of hunting takes place in this area. We turned left off the road and walked about a hundred yards through the woods to find a sand bottomed ditch with a shallow trickle of water. When John was here last, the creek contained two feet of flowing water. We turned back to the road and walked toward a large beaver pond that we had seen when we first entered the forest. John led us through a tangle of small trees and brier bushes to a large beaver dam. The dam was old and pocked with holes that sucked water in spiraling swirls only to emerge into one of the streams on the lower side of the dam. Most of the vegetation was dead at this time but in a couple of months, leafy trees would block any passage here. We also found the feathers of a dead great blue heron along the pond side of the dam.

We headed back to the Goose Nest Stream to explore it. We climbed down the slick red clay bank to a sand bar and took pictures. Some of the guys walked down the stream to find the Baker's Pond Stream. The Goose Nest Stream had twice the flow of water and a higher altitude. It is more likely to be the true source of the Wolf.

From what we had seen so far, it is obvious that the Wolf springs from many different sources. According to the national forest sign, Baker's Pond was just on of them. John said that the locals call this whole spring bed Baker's Pond. I could see why. We were standing in the bottom of a spring filled valley surrounded by hills of red loess. I tried to imagine what it would have looked like before settlement. A bottom area comprised of beaver ponds and streams. It might have appeared as one large body of water.

On the way back, we stopped at the house of someone who lives near Black jack Road. He gave us a tour of his land and even offered to let us camp there in the future. He explained that the government cut a canal to the right side of the river to straighten it out for barge traffic. Starting in the early 1900s, barges would carry grain from this county to Memphis. In the 1940s, trucks had replaced barges for shipping, so the canal was ignored and allowed to silt in. Even though it appears as a straight blue line on some maps, it is now a grassy field. The landowner also told us stories about the otters, snakes, beaver, deer and people that inhabit this area. Everyone we met was amazingly welcoming. It is refreshing to meet people that are genuine and nice.

RIVER ANGELS

To end the day, we stopped at the Black Jack Road Bridge over the Wolf to take pictures and video of the setting sun. The sun was only a glow on the horizon as a beaver crossed the dark water to where we were standing. It saw John and turned back. However, it seemed to forget there were people on the right bank because it circled back only to turn around again. The beaver continued to circle until we left.

On February 27, 2012 at 11:35AM, John Hyde wrote:

Rod, Any luck on the hammock sponsorship from Welty? We scouted the exact launch point for SUPAS yesterday and we'll sure be doing at least a few wild camp sites. Hammocks would sure come in handy - as at least one site may be inhospitable to tent use... Thanks, and I look forward to this adventure! :-)

Thanks John

On February 27, 2012 at 1:32PM, Keith Kirkland wrote:

Thanks John Ruskey – and everyone else who is working to bring attention to the beauty and adventure that can be found on our Wolf River!

Thanks so much, everyone, and have a great time and a safe trip!

Thanks,

Keith Kirkland
Wolf River Conservancy
Director of Outreach Lands

On February 28, 2012 at 9:44AM, Dale Sanders wrote:

Thank you Jonathan & John. Just for info. Subject visit was a very productive. I know now, it is possible to paddle/portage, the Wolf River from it's source to the Mississippi. Not a lot of paddling though in the upper sections. Advise everyone to get in shape for dragging craft over the most

God Forsaking terrane I have ever experienced. An under statement - "It will be a challenge". I am, now more than ever, looking forward toi the trip. Certainly would be a great Navy Seals training area. Almost every day now, trip good news info is coming out; believe Jonathan & John have arranged it so we won't have to drag our camping gear through the first day swamps. Hurray! Very Respectfully, F. Dale Sanders

On February 28, 2012 at 2:50PM, Jonathan Brown wrote:

I fully agree with Dale! After looking at where we will be starting the trip and remembering what it was like for me 6 years ago, it will be super important to be in top notch shape for this thing.

For those of you that didn't know, this past weekend, Dale, John, Richard, Ray Graham, and Mike Watson went down to the area the Wold River starts. We trekked to our potential starting points and looked over Baker's Pond. We also met with one of the landowners, James Lowrey.

The outcome of the trip this past weekend was this:

- The first three days will be no picnic.

- We will fully need a logistics team to bring and take gear so we have as little as possible.

- We can camp on James Lowrey's property.

- We got the names and number of the local Game Wardens.

- We learned that land owners we were concerned about may not be an issue.

There was one line in Jonathan's email that alluded to a subject that, up to this point, hadn't been openly discussed in group correspondence: "*…remembering what it was like for me 6 years ago…*"

During a recent Skype conversation, Jonathan touched on it briefly, but refused to discuss it in any detail, writing it off as the insignificant wanderings of a younger, more naïve version of himself. The subject would remain a mystery until I arrived in Memphis in early April to start the Wolf descent. The full story, however, wouldn't be shared until I interviewed him in July 2015 during downtime from yet another paddling excursion.

As a child, Jonathan Brown was on the move a lot. His family lived in Tennessee, Wyoming, and, on two separate occasions, Hawaii. A move back to Memphis in his teens landed him close to the Wolf River. The river provided JB and his high school friends with a place to exercise their inner Huck Finns, albeit *their* journeys were supplemented with techno music, pool rafts, bottles of IBC root beer, and, of course, a healthy dose of naivety. During one nighttime river adventure, JB and his friends inadvertently ran into a snag.

"We came over a weir that we did not know was gonna happen, and we saw it and we had to go over it, and all the pool rafts completely opened up—you know, broke—and we barely made it to the shore saving everything. It was horrible. We started at about 11pm and ended at about two in the morning. So…*fail*! But, that's what sparked the idea that we should paddle from the beginning of the Wolf River to the Mississippi River."

JB and his friend Kevin Brunson secured some topographic maps of the upper Wolf and began plotting their course. Their maps showed an abundance of water in the headwaters area, but, as they would soon discover, the difference between what's on a map and what really exists are two completely separate truths.

They rented a large tandem canoe from Outdoors Inc. (a local sporting goods store and supporter of SUPAS), packed three days of food into the boat, and drove across the state line to Mississippi.

The easiest access to the river was the bridge at Chapman Road, about four miles downstream from Baker's Pond. (They didn't know at the time that the area around Baker's Pond was the true headwaters.) Kevin's girlfriend, Angela, dropped the pair at the bridge and had some concerns about their safety, especially when they learned there was no river running under the bridge.

"It was like this little creek," says JB, "this super small little flowing thing of water that the canoe didn't even fit in."

It never occurred to them to drive out in advance and scout the area.

"Basically, we went in blind."

Despite the obvious obstacles straight out of the gate, and despite Angela's concern, JB and Kevin remained highly optimistic.

"She was like, 'I don't know guys, this doesn't look good. But we were like, 'No, no. If you look down through those trees, it'll open up. The river's like right there.' We told her, 'We'll be good. We got everything. We got a machete. We got a BB gun. We're good to go!'"

To their credit, they had size on their side.

"Kevin's a big, ol' 6'3" guy," said JB. "He had been in the Air Force academy already. He had gone on ROTC (Reserve Officers Training Corps) throughout all high school. I mean, he had the gear, the stuff. He was ready to rock it."

"But did you guys have experience in a boat?" I asked.

"He had more than me, but neither of us did," answered JB. "But we had that sense of *adventure*, that idea of the unknown."

"And a bit naïve…?" I prodded.

"Oh, what do you mean a *bit*?" said JB, chuckling. "It was *100%* naïve!"

Obviously, some things hadn't changed since high school.

Undaunted, the pair set off downstream, hoping to arrive at the Mississippi River in three days.

"It's July. It's *super* hot. *Super* humid. And it's, like, *snake season*. We got in the boat. Couldn't float. We started pullin' it. Whenever we found a little pocket of water, we'd float it, and then we'd get back in. And we're all good to go. We're like, 'Sweet, we're gonna make this happen.'

We're goin' down. We're in the woods. And then we come out and it's just this open kind of marshland with dead-looking marsh trees just kind of stickin' out everywhere and just nothin' but sawgrass, with water weaving its way between all the sawgrass. In a lot of spots it was deep enough to both get in, but because the canoe was so large, it was hard to navigate around the sawgrass. And when it wasn't deep enough, we had to go *through* the sawgrass."

It quickly became apparent to JB and Kevin that they'd gotten in over their heads, so to speak.

"We weren't even prepared dress-wise. Kevin's got, like, his camo gear on. I've got, like, I think, *board shorts* on, and flip-flops. I don't even have shoes. Because, who even thought we'd be getting out of the canoe? We'd just be on a river the whole time."

JB shakes his head and continues.

"And so, we're havin' to tromp through the sawgrass, which is just cutting our legs up left and right—just cutting from the minute we started this. And we were like, 'This is insane!' And we're just thinkin', 'It's gonna eventually become a river! Like, this is the water to the river. It's gonna get there. This has gotta be it!'"

On the edges of the swamp, JB and Kevin could hear cars speeding along a distant road. The sounds were comforting and they knew they could hike to the road if things got dangerous. Begrudgingly, the pair trudged on through chest-deep water, snake nests, and vicious sawgrass.

"It was nothing but marsh the entire time—marsh and sawgrass. So, we ended up saying, 'The sun is starting to go down and we don't know where we are. We need to get somewhere.'"

Deeper into the swamp they plodded, until finally they had to make a critical decision.

"We notice we could go left and go further in the marsh, or we could go right—where there's some trees. So, we went right and we noticed there's an old, dried up riverbed and we're like, 'Ohhh, this might be...*some*thing! Let's follow this. This has to go *some*where.'"

The heat, humidity, and physical exertion was taking its toll on Kevin.

"He had just been worn *out*. The heat had gotten to him, so he was starting to feel really faint. We have a machete, so I'm up front just trying to cut our way through everything. We're just trying to drag the canoe over trees and limbs—nothing's wet, it's all dry. There was cobwebs and spiders all up in our canoe, crawling on us. We're getting bitten by this stuff. We're goin' against poison oak, poison ivy. I mean there wasn't one thing—except for the snakes—that was not biting us or making us itch. Our legs were bleeding. Our arms were bleeding. It wasn't like dripping blood, but we saw all the lines and red was coming out of them, all the way down to my feet—because I didn't have any shoes. And Kevin, because he was so heavy, the mud was going up to his *knees*."

They soldiered on and finally heard cars on a nearby road. The clearest route to the road would take them through a forested area where the ground was completely covered in sawgrass. They weighed their options before proceeding.

"I said, 'Dude, the sun is setting. It's getting dark. We cannot be here. The mosquitoes are already out. We're getting bit up. If we're here now, and it's dark, we're done. Not done like *dead*, but this could be pretty dangerous.' So, we decide to leave the canoe—because we can't pull it anymore—and mark the trees and work our way out."

They made it to a rough pathway on solid ground and noticed a bluff in the distance.

"At the top of the bluff there's a trailer, truck, and everything and we're like, 'Okay, do we go up to the door?' And I'm like, 'Kevin, you're *huge*. We're *bleeding*. This is not the best scenario.'"

Because he was the smaller of the two, and wasn't draped in blood-soaked combat clothes, Jonathan decided to approach the trailer door.

A guy in his mid-20s answered. His name, ironically, was also Jonathan. He lived in the trailer with his wife and kids.

"He was a real southern bumpkin guy. He said, '*Where'd y'all start?!?*' We told him, 'The bridge. About three miles upstream.' And he's all like, '*Man!* Y'all came through the marsh and the *trees?!* Our family's owned this property for about 20 sumthin years and I don't know of *any*body…I ain't never even been in that area! We don't know how to go *through* that area!'"

Besides being dumbfounded by the fact that two blood-soaked paddlers had just arrived at his doorstep, landowner Jonathan was glad they were safe and happy to help them in any way he could.

"They just bent over backwards for us," says JB with a smile.

JB, Kevin, and Trailer Jon drove down to the bottom of the bluff and, with some sincere effort, retrieved the canoe from the swamp. JB and Kevin were given soap and towels and they proceeded to hose themselves off alongside the trailer.

After they got cleaned up, they used Trailer Jon's phone to call Kevin's girlfriend and waited in air-conditioned comfort while she drove out to pick them up. Unfortunately, Angela's arrival opened some new wounds.

"When she rolled up, we weren't thinking, and we got up real fast. All of our wounds, with the air-conditioning, had started sealing. When we got up, every wound opened and we both yelled. She (Angela) took us by a Walgreen's and we got the biggest things of Neosporin and hydrogen peroxide that you could possibly buy. She dropped me off at my apartment. I jumped in my claw foot tub, laid down butt naked, took the whole bottle of hydrogen peroxide, and poured it over my entire body. My whole body foamed. My arms were shaking. My legs were shaking. My whole body was shaking, just by the nature of pain. I took all the Neosporin and covered my legs, and covered my arms, and just laid out.

For the next three weeks, Kevin and I had to wear long sleeved shirts and pants because it looked so bad. Our fingernails and toenails turned yellow. All the poison ivy, poison oak that was on our bodies was *thick*."

JB paused to smile before launching his punchline.

"And then we said we would do it again!"

Years later, JB, Kevin, and Angela returned to northern Mississippi to properly thank Trailer Jon and his family for their kindness. Their visit included some sad news, but it also produced a key connection that would later benefit the SUPAS expedition.

"We knocked on the trailer. Nobody showed up. But there was a new trailer next to it. It turned out to be Jonathan's aunt and uncle. We found out that just years prior to that, Jonathan died. He had a drug overdose in that trailer. His family was still living in that area. When Jonathan died, nobody could foot the bill for the property, so the gentleman who lived in a newly built house—*behind* the trailers—bought the whole property."

That gentleman's name was James Lowry. As a favour to Trailer

Jon's family, Lowry allowed them to continue to live in the area they'd called home for over 30 years. Not only had Trailer Jon been a "river angel" to JB and Kevin on that fateful day many years ago, but now, years later, and on the same piece of ground, James Lowry was cementing his own role as a "river angel," not just for Trailer Jon's family, but for SUPAS as well.

With the impetus of a Wolf River descent brewing in JB's brain, a unique series of events began to unfold, paving the way for the birth of SUPAS.

The first event was JB and Kevin's revisit to James Lowry's property in the summer of 2011. During the visit, Lowry gave them a tour of the property. To gain better access to the river channel (the one JB and Kevin had inadvertently missed during their day in the swamp years prior), Lowry had cleared a huge swath of trees and sawgrass. The location of the channel—now plainly visible from a dirt path at the swamp's edge—and its proximity to Lowry's house lodged itself in JB's head that day. He would pull on that memory months later when the SUPAS scouting crew went looking for possible places to camp during their expedition.

In early 2012, Lowry, ever the generous host, graciously allowed the SUPAS team to use his property as a home base for their Day 1 camp. Having a secure place to store their gear and park their shuttle vehicles greatly eased logistical woes for the team. Because property adjacent to the upper Wolf was private on both banks, communication with the landowners—initiated by John Hyde—had been necessary in order to establish camps for the team. And because the expedition was being promoted in all forms of media in Memphis, the SUPAS organizers wanted to be seen in a positive light, not as a bunch of squatters and trespassers. Although they didn't know it at the time, JB and Kevin's tortuous 12-hour trek to Trailer Jon's property all those years ago eventually helped pave the way for a historic descent of the Wolf River.

While living in Hawaii in his 20s, JB developed a love for surfing. When he moved back to Memphis in 2009, surfing wasn't possible, but stand-up paddleboarding certainly was, especially on the Wolf.

Unfortunately, he didn't have enough money to buy a SUP. One day, he found a windsurfing board on the side of a road and decided that he'd try to paddle it like a SUP. He fashioned a paddle out of a tree limb and hit the Wolf with big aspirations.

Disappointingly, the board and tree limb didn't perform as expected. But fortunately, during an afternoon of downriver "paddling," he met Ray Graham, a Wolf River Conservancy river guide. During conversation, JB mentioned his desire to paddle the entire Wolf from end to end, albeit on a proper SUP. Ray, whose passion for the Wolf spans a lifetime, liked JB's idea and suggested he get in touch with "The River Rat," aka Dale Sanders, another WRC river guide.

"Dale has been wanting to paddle the whole river for years," said Ray. "Give him a call. He'd love to hear from you!"

A short time later, a grey-bearded man with bright eyes and a friendly demeanour walked into the Apple store where JB worked and asked for some computer assistance. JB, always quick to help a stranger in need, shook hands with the man and the two of them got down to work. That man, as it turns out, was Dale Sanders.

Weeks later, JB was out for a leisurely walk at Memphis' Shelby Farms (one of the largest urban parks in North America) when he spied a short row of SUPs leaning against a small kiosk at the shoreline of one of the park's lakes. The sign on the kiosk read: SUP Memphis. It was exactly what JB had been looking for—a stand-up paddleboard rental business. Ironically, the shop had opened that very day. JB and SUP Memphis owner, Luke Short, began talking about surfing, SUPing, and JB's idea to paddle the entire Wolf. JB explained that his current craft was an old windsurf board that he'd been paddling with a tree limb. Luke admired JB's passion and decided to give him a paddleboard at no cost. Needless to say, a deep friendship formed that day.

Luke Short then helped spread the word about SUPAS in spring 2012 and joined the team on Day 1 of the expedition. Luke got lucky—Day 1 was one of the most difficult days of the entire trip. His introduction to the upper Wolf was unforgettable.

Another significant SUPAS connection revealed itself in the

summer of 2011. Through SUP-related websites and social media pages, JB learned of Dave Cornthwaite's descent of the Mississippi River. He contacted Dave directly and planned to meet with him when Corn's downstream journey arrived in Memphis. Unbeknownst to Dave was the fact that JB was organizing a full-blown Southern welcome, complete with TV cameras and a flotilla of kayakers, canoers, and SUPers. Members of the Wolf River Conservancy and Bluff City Canoe Club joined in for the 17-mile paddle from the Shelby Forest boat ramp to Mud Island Marina near the foot of Beale Street. Future team members of SUPAS were also amongst the paddlers. The flotilla took up a sizeable collection and donated the money to CoppaFeel!, a breast cancer awareness organization that Dave was supporting on this expedition.

During Dave's departure from Memphis, JB shared his plans to paddle the Wolf the following spring. Corn showed sincere interest in participating in the event and later committed fully. It was also on this day that JB and I met for the first time.

With Dale Sanders, Ray Graham, and possibly Dave Cornthwaite now interested in the Wolf descent, JB contacted his old friend and fellow missionary, Rachel Sumner, in November 2011. (The pair had been involved with the local and overseas missionary groups Youth With a Mission and University of the Nations.) Rachel, who had helped raise awareness of Dave's Mississippi journey trip by coordinating local media events in Memphis, was also one of the founders of Operation Broken Silence, a Memphis organization bringing awareness to sex slavery and human trafficking in the Memphis area. When JB shared his idea of paddling the Wolf River with others, Rachel was more than enthused to get involved. They agreed that money raised from the SUPAS expedition would go toward the building of a recovery house for survivors of human trafficking.

With all of the necessary components now in place, Stand Up Against Slavery (SUPAS) was born. From its inception to its maturation, JB had been the decisively driving force behind the project. SUPAS mirrored his love of helping people and bringing them together. Surely, the title of "River Angel" is not handed out generously.

RIVER ANGELS

It is earned through countless hours of selfless work, work that doesn't often feel like work because it comes straight from the heart. Jonathan Brown, Rachel Sumner, Luke Short, Dave Cornthwaite, Ray Graham, Dale Sanders, and James Lowry all give of themselves in ways that definitely qualify them for "River Angel" status. And yet, at the end of the day, prominence—a word all of them would shirk at—has little significance in a world where equality and altruism proves far more beneficial to the masses than greed and disparity. To quote the Tao Te Ching, "The heart that gives, gathers."

Meanwhile, on the Bikecar front, a major decision had been reached.

On February 26, 2012 at 12:37PM, Dave Cornthwaite wrote:

Hello buddy,

Ok, Bikecar expedition is ON. Let's have a chat this week about the best way to do this/ get bikecar to Memphis (via you, or otherwise) and then post Memphis-Miami what to do with it.

Here's my page - will add you into the write up shortly my friend www.davecornthwaite.com/#/2012-bikecar-memphis-miami/4561260435

Dave Cornthwaite

Corn's good news arrived just as I was packing the last of my worldly possessions into a storage locker in Vancouver. I'd given notice on my apartment and was vacating it by week's end. I would then board a plane to Ontario. Unbeknownst to me, it would be another two-and-a-half years before I set foot in Vancouver again.

Despite the good news from Corn, this little note from Jeff Lozar in Eugene got me riled. It seemed our bike-friendly friend, Paul Adkins, had yet to follow through on his promise to retrieve the Bikecar.

On February 28, 2012 at 9:57AM, Jeff Lozar wrote:

Rod,

Just a heads up the bike car is still sitting in my storage lot. I have received one email from Paul a couple of weeks ago which I replied to, other than that all is quiet here. Have a great day today!

"Why can't people just do what they say they will do?" I said aloud through clenched teeth. "Fuck!"

And with that barely stifled outburst, I said goodbye to the west coast and flew directly to Ontario. The next day, I sent an email to Corn.

Dave,

Sorry for the delay, mate. Just got to Ontario tonight. Been crazy busy this past week.

That's good news about the bikecar. Received an email Monday from Jeff Lozar, the guy who has been storing the bikecar in Eugene since last autumn. Apparently the new storage guy, Paul Adkins, has not picked up the bike yet. I will call Paul A on Wed to find out what's happening. I was hoping he had moved it and had a chance to examine it so we could figure out the best way to break it down and ship it. But...he has yet to see it. I'm hoping he can move it this week/weekend. If not, Jeff Lozar will be asking for another $50 for March's storage fee. I sent an email to Jeff today ensuring him that we have April plans for it and that it will be removed from his property very soon.

Chatted with Paul Everitt about breaking the bike down. He said everything can be removed from the frame but the frame cannot be broken down. As you can see from the video posted on your Bikecar page, the frame is fairly big, and heavy. The bike is 100kg sans gear. Shipping it could get expensive. It might be possible to build a box out of plywood, remove the wheels and steering column (to reduce the size) and ship it that way. Dimensions: 1.7m wide, 2.6m long, 1.4m high.

I will call some U.S. shipping companies on Wed to get quotes. It would be good to chat wth you about this. I'm available all week. Just need to set up a time that works best for you.

Cheers,

Rod

(P.S. Saw Date book tonight. Lookin' good! Well done!)

On March 2, 2012 at 3:05AM, Rod Wellington wrote:

Dave,

Re. Skype chat: Sat/Sun/Mon are all good for me. I will be at home working on the computer almost all weekend. (Although, there will be a couple bike rides thrown in too!) So hooking up should not be a problem. You are 5 hours ahead of me - Noon in London is 7am here. So, I'm thinking a 2pm London - 9am Ontario meeting any day could work. You mentioned nights were out. Afternoon or morning is fine. Let me know if the above time works for you and which day is best. I'm available all three days.

Haven't heard from Paul Adkins in Eugene yet. Will try to call him again on Friday. He has no phone voicemail, so can't leave a message. Sent him an email but no response yet. Working on ideas for shipping the Bikecar. Consulting with people here and getting some good feedback. May have some simple ideas that will work. There is a European company that sells quad bikes. I will email them and see how they ship their bikes.

First leg out of Memphis is a go. I can line up a return (car) ride back to Memphis. Perhaps Dale or Richard can be of some assistance. Also, what do think of me picking you up at the Memphis airport with the Bikecar? Fun photo op/video op.

Cheers,

Rod

On March 2, 2012 at 7:50AM, Jeff Lozar wrote:

Paul/Rod,

I should be in the office most of today, call me to set a time to pick up. I thought about calling you but I tend to avoid making calls this early unless I know you are an early riser. The storage fee for March was due yesterday ($50). I will waive the fee if you get it picked up today as I would love to have this finished.

Thanks,

Jeff Lozar

From the tone of his email, I got the impression Lozar had long tired of our Bikecar shenanigans. We had to act *fast*.

March 3, 2012 8:45AM via TweetDeck

DaveCorn Dave Cornthwaite
Making expedition plans with Mr Zero Emissions Expeditions, @rod_wellington who will be joining me out of Memphis on the #bikecar

Before going to bed, I sent an email to the SUPAS crew in Memphis, notifying them of my post-SUPAS plans.

Gentlemen,

Happy to say that I'll be joining Dave on the Bikecar for the first few days out of Memphis. We don't have a departure date yet. We will leave when our Memphis business is finished, of course.

I will need to line up a ride back from a city south of Memphis, perhaps a three-four day pedal (120-150 miles south of Memphis. Our plan is not to rush, maybe 20-30 miles per day.) So, need to find out about the possibility of getting picked up 3-4 hours south of Memphis and getting back to Memphis by car.

Dale - Dave and I need to ship the Bikecar from Eugene, Oregon to Memphis. Can we arrange to have it delivered to your house? I can arrange it so that it will arrive after I arrive in Memphis, so between April 3-6. I want to have a look at it before we leave for the Wolf. Also thinking that it would be funny to pick up Dave at the airport with the Bikecar. Would make a great photo op/video op.

Cheers,

Rod

On March 4, 2012 at 8:49AM, Dale Sanders wrote:

Rod

We will get you picked up down the road after biking the car. HA! Lets see what Jonathan comes up with first though - save me if no one else comes forth. Please feel free to ship the bile car to my home, anytime is OK (we have plenty of inside storage). However the photo ops at airport really sounds neat. I will also be available to pick Dave up at the Airport, if needed. Lets do what's best for marketing - Jonathan may wish to have some official from OBS pick him up. Either way is OK with me.

On March 4, 2012 at 3:11PM, Rod Wellington wrote:

Dave,

Some info below from Dale regarding walking distance in the first two days on the Wolf.

Also, Dale confirmed that he can receive and store the Bikecar at his place.

I'm thinking that it would be good to ship it ASAP, perhaps within the next 7-10 days. I will have a better idea of a timeframe when I list the bike on UShip. I need to know how we will pay for the shipping. You mentioned that you can cover the cost. What are your thoughts on making payment? At this point, not sure of UShip's payment arrangements. I will be able to send you the final auction/bid quote and you can take it from there. Also thinking that you may be on the sailboat by the time the Bike-car is shipped and how that may complicate things. Would it be easier for me to cover the shipping and you reimburse me?

Cheers,

Rod

On March 4, 2012 at 3:46PM, Dave Cornthwaite wrote:

Good to see this Wolf info, and also to know that Dale will receive the bikecar.

From Wednesday my access to Internet will decrease and then on the 14th I'll disappear until April. If you're happy to cover the cost and for me to reimburse you then that might be the easiest option. Agree that shipping it ASAP is the best option.

Cheers matey

Dave

On March 4, 2012 at 5:32PM, Paul Adkins wrote:

I can give you a call this afternoon. I have the Bike Car and it is at my place.

"Hallelujah!" I shouted.

Corn probably heard my celebratory howl all the way across the Atlantic, but I sent him an email nonetheless.

RIVER ANGELS

Hi Dave,

Heard from Paul Adkins in Eugene today. (We spoke on the phone.) He has picked up the Bikecar and is now storing it at his house. He pedalled it there. Said it was straightforward and fairly easy to pedal, though heavy. Said he had to lift it over a short fence by himself at his place but was able to do it - front wheels, then rear wheels. He estimated that it weighed approx 150lbs. (vs. the 100kg - 220lb quote I got from Paul Everitt). Regardless of the weight, the fact that he was able to muscle it around by himself is a good sign. He also said that a tube was missing from one wheel - he had to pedal it flat, said that he would insert a new tube in it. Also, there are about 4-6 spare wheels and tires, all of which are now in the big plastic/fibreglass storge box on the back of the bike. Paul E had also disconnected the rear brakes and had removed the chain running from the passenger pedals/cranks to the rear axle. That chain is in the storage box. So that stuff will need to be replaced/fixed in Memphis. He said that he could easily pedal the bike to a shipping yard/business in Eugene if needed. I told him I would try to arrange it so that a truck would pick it up at his house. Re. pick-up: He gave me a time of day that works best with his work schedule. Told him that I would do my best to have the bike picked up by the 15th or ASAP. I am planning to list it on UShip tomorrow.

Things are moving along...

Rod

On March 5, 2012 at 4:34AM, Dave Cornthwaite wrote:

Beautiful. Great news all round, thank you kindly, Sir....

  Dave and I drafted up an advert and posted it on UShip.com, an online auction platform for shipping items in North America, especially big ticket items that need to be moved via transport truck. Independent truck owners and businesses bid on your item's shipping and you choose the rate that suits you best. It seemed like an extremely popular service, judging by the fact that it had spawned a television reality show (Shipping Wars). It also seemed like an efficient and cheap solution to our shipping situation. We crossed our fingers and hoped for the best.

By week's end, the UShip advert (below) had received 13 quotes ranging from $559.27 to $7236.83. Ten of the quotes were under $1000. I was relieved to see that.

Four-wheeled bicycle (Bikecar)
- Length: 8' 6"
- Width: 5' 7"
- Height: 4' 7"
- Weight: 220 lbs

Additional Information: This 4-wheeled bicycle (Bikecar) is to be used for a 1001 mile charity ride between Memphis and Miami, by British Adventurer Dave Cornthwaite in April and May 2012. His journey will be raising funds for breast cancer awareness.

Dave's website and more information about the journey can be found at (N.B. Website URL was removed by UShip.).

The Bikecar is to be shipped as is, as you see it in the pictures, minus the flags and the bearded gent. All loose items have been securely stored in the rear box affixed to the bike.

As stated above, the Bikecar journey from Memphis to Miami is a charity fundraiser. We are looking to keep costs to a minimum. Discounted or donated free carrier service is most welcome. In exchange for discounted or donated free carrier service we can offer prominent placement of the carrier's name, logo and website link on Dave Cornthwaite's website and in social media prior to and during the Memphis to Miami Bikecar journey. This is a great opportunity to get free publicity for your shipping company.

We are looking to get the Bikecar Memphis-bound any time after March 13.

RIVER ANGELS

Meanwhile, in Memphis, there was some trouble brewing between John Hyde and Dale Sanders. Judging by the tone of his emails and his insistent mannerisms during the Skype meeting the previous week at Dale's house, John seemed like a headstrong and hot-headed individual (with emphasis on *individual*). Interestingly, my thoughts about Dale Sanders were the same. Dale's opposition to John's claim that Baker's Pond was not the true source of the Wolf River was obvious to everyone involved with SUPAS. The camp seemed split on who was right. Leaning on hard, provable evidence, John claimed that Goose Nest Lake (located a mile east of Baker's Pond) was the true source. Problem was, the lake was on private property and was not an ideal spot to launch an expedition. The Wolf River Conservancy—an organization with which many of SUPAS' members belonged, including Dale—claimed that Baker's Pond was the true source. A few feathers had been ruffled by John's claim, and, whether purposefully or not, he'd gotten under the collective thick skin of his teammates, particularly Dale.

For reasons unknown, John chose to leave the meeting early that evening. With John gone, Dale seized the opportunity to openly voice his suspicion and displeasure of John. As Dale's rant unfolded, I sat at the other end of our group Skype conversation feeling awkward, appalled, and confused.

Up to this point, I'd trusted Dale. His positive support of SUPAS through email correspondence had won me over. He'd gone out of his way to ensure that Cornthwaite would be present for the Wolf River descent. He'd promoted both my and Corn's causes online and in emails. And he'd offered to house my father, sister, and brother-in-law during their stay in Memphis. But his open display of back-stabbing on this night made me nervous. My trust was now waning, and I knew it was just a matter of time before an eruption between Dale and John ensued.

On March 1, 2012 at 11:49PM, John Hyde wrote:

Dale - could you ask Mary (Finley) *that if she's gonna forward that email, to change my name to John Henry and DROP the Hyde. I thought you understood that the main reason I wanted to be associated with the*

SUPAS trip was to promote my Art! I explained to the whole truck full of six (including you) who were @ Baker's Pond Sunday that it's easier to sell Art photo prints as John Henry. I go by John Henry artsitically, and Henry is my legal middle name. I have furnished Jonathon with many photos for promotional purposes on the website, and he is posting my Facebook Page's email as the link to my Art. My Facebook link is "Facebook.com/ JohnHenryPhotography". PLEASE correct this! I am not ashamed of my last name, BUT - as an artist - I go by John Henry.

Thank you - respectfully - John Henry Hyde

And to further drive the point home, this less-than-friendly correspondence—addressed to Dale, but forwarded to all SUPAS members—arrived in my inbox the next day:

On March 2, 2012 at 12:22PM, John Hyde wrote:

Good morning Dale - I am doing SUPAS from the headwaters area down to new 72 because Keith Kirkland told me about the trip and asked if I wanted to be involved) (I assume this was because of my previous experience in the Headwaters Area of the Wolf). I am furnishing images to SUPAS for their website promo link because Jonathon asked me to. As to selling images - I do NOT actively market them. I have sold less than 10 in my lifetime. As to my name - I wish to be know as John Henry for the purpose of promoting my Art, and my Facebook Page : John Henry Photography only. I am NOT a sponsor. I have already spent well over $100.00 ,out of MY own pocket, on gas alone, not to mention 4 trips to the headwaters area this year alone, and will NOT be sponsoring anything. Jonathan, Rachel, AND Robinson are all aware of my request to be referred to as John Henry , as far as any press releases or promos are concerned. The link to my facebook Page was ALL that I have asked of OBS/SUPAS, Keith Kirkland, or ANYBODY else associated with this EPIC adventure, for my time, money, and considerable effort in making SUPAS a success.

Honestly - I was surprised and insulted to be listed as a "section paddler" in your email to BCCC. I am a SUPAS team member (whether I do the whole trip or not). I have openly expressed recently that I intend to boat down to new 72, and have offered not only my 3-man boat, but also my time on the legs downstream of New 72. I am sure you remember my offer to be "Plan B" if SUPAS members don't make it to St George's on time. I also planned to paddle at least Bateman Rd to Moscow, AND the leg which ends up @ the mouth of the Wolf. If anything - I am advance scout/guide for the SUPAS trip, NOT a "section paddler". I have done MORE to scout that area for SUPAS than ANYONE ELSE, AND am still the only one involved who has already boated from Blackjack Rd to Memphis.

If you do not want me to continue with my efforts (including a scheduled scouting trip Sunday in order to locate/secure a place for camp 2 of SUPAS) just say the word, and I will bow out gracefully right now.

I am NOT a money-grubbing Photo-hawker, BUT - if I cannot at least be represented the way I want in emails concerning SUPAS (whether they be to BCCC members, WRC members, or anyone who logs onto the SUPAS website link - perhaps it would be best if I left SUPAS to you, and the "Team Members" you listed in your email...

Respectfully - JOHN HENRY Hyde

The following emails were *not* sent to John Henry Hyde. He was excluded from the conversation that pertained exclusively to him. Emails were addressed to Jonathan Brown, Mary Finley, Robinson Littrell, Dale Sanders, Luke Short, Richard Sojourner, Rachel Sumner, Mike Watson, and myself. Their content made me feel even more awkward about the situation with John *Henry* Hyde.

On March 2, 2012 at 8:08PM, Dave Cornthwaite wrote:

Dear all,

I must admit, this last email from John seems a bit over the top. Regardless of experience, any cracks showing in team spirit over such petty things are going to damage the success and enjoyment of the journey. This email was a big red flag to me.

Thoughts?

Dave

On March 2, 2012 at 9:42PM, Richard Sojourner wrote:

Dave, I agree there has to be something else going on for John to be so sensitive!

On March 2, 2012 at 9:56PM, Mike Watson wrote:

Those who are not flexible, break. :-) I have only met him once and he seemed nice enough. But as I told Dale on the phone, it was a little curious getting an email from a JohnHyde wanting to be referred to as John Henry.

JohnHenryArt and JohnHenryPhoto are both available at AOL if anyone cares to let him know.

On March 2, 2012 at 10:28PM, Robinson Littrell wrote:

I'm not sure what's going on, but if things are being said/announced about SUPAS that are not going through either Rachel, or me or even Jonathan Brown, that is not good. We at OBS (who are hosting SUPAS) respectfully ask that any communications with a 3rd party outside the SUPAS team and OBS (including Email blasts and press releases originating from SUPAS team members) be reviewed and checked over by OBS staff and/or Jonathan Brown. While we greatly (and I mean GREATLY) appreciate your enthusiasm and desire to spread the word and make this event incredible, I'm sure you all can understand that we want as few variations from the actual event information as possible (though we know that they are unintentional mistakes). It helps prevent confusion further down the road.

As for the team, if there are any issues between individuals, please sort those out privately as opposed to letting everyone involved with SUPAS know.

John Hyde will from now on be referred to as John Henry officially.

Thank you all for your involvement and I'm very excited to see what this will turn out to be.

God Bless,

On March 2, 2012 at 10:35PM, Jonathan Brown wrote:

Hey SUPAS friends!!!! Thanks for all the open communication. I emailed Dale and John separately. Miscommunications, misunderstanding, and how they are handled is I walked away with in my head after reading the string of emails. And to boot, communication via email never proves to be a trusted friend when wanting to resolve things. :)

Personally I think it would be beneficial for Dale and John to connect aside from email to iron out any wrinkles.

That was my two cents.... now I'm broke.

Jonathan

Early the next morning, John Henry was being resurrected—at least by some.

On March 3, 2012 at 9:29AM, Mike Watson wrote:

I'm working on a map for the Wolf River Expedition in the near future and thought the attached was interesting. It supports what John Henry has been saying about the source.

Mike

Attached to Mike Watson's email was a Google Maps satellite image screenshot of the area around Baker's Pond, the alleged source of the Wolf River in northern Mississippi. Mike had highlighted a low ridgeline that ran along Highway 72 just north of Baker's Pond. The ridgeline represented the furthest edge of the Wolf River drainage area. Any precipitation that fell on the north side of the ridgeline would end up in a different drainage system. Mike had also highlighted Goose Nest Lake and the lake's outflow, Goose Nest Creek. Both were about one mile west of Baker's Pond. Goose Nest Lake was located at a higher elevation than Baker's Pond and it was located further from the Wolf's mouth than Baker's Pond. (Elevation and distance from the river's mouth are two criteria taken into account when determining a river's source.) Mike's map lent positive credence to John Henry's claim that Goose Nest Lake was the true source of the Wolf River. Although John was rocking the collective SUPAS boat with name changes and boldly worded emails, he was also being vindicated based on his knowledgeable contributions in finding the river's utmost source.

On March 3, 2012 at 1:18PM, Jonathan Brown wrote:

Dude that's exactly what John (Henry) was saying. Pretty cool finding indeed! Mike you are amazing for putting this map together and John you are amazing for jumping into the deep end researching all this!

Jonathan

On March 3, 2012 at 2:18PM, Ray Graham wrote:

That's great, what a nice find. Here is the youtube site where I posted some videos from last week. I'll put some more up later.

www.youtube.com/user/wolfriver100

Here is the blog: www.wolfriverguide.blogspot.com

RIVER ANGELS

On March 3, 2012 at 2:37PM, Rod Wellington wrote:

Mike,

Thanks so much for putting this map together! It's great to see a bird's eye view of the Baker's Pond area and the route of the infant tributaries. And thanks to those who scouted this area and accumulated the data firsthand.

Cheers,

Rod

On March 3, 2012 at 7:25PM, Rod Wellington wrote:

Gents,

It would be good to know the intended walk distances of the first one or two days. I am assuming that we will be on foot for at least the first day. Correct? And that we will rendezvous with our watercraft and gear somewhere downstream after the first day. Correct? Have you determined where that rendezvous will be and how far the walk(s)/portage will be?

I'd like to get a breakdown of what those two or three days are looking like at this time.

Cheers,

Rod

On March 3, 2012 at 8:31PM, Jonathan Brown wrote:

What I do know is that we will be floating, dragging, and carrying our watercraft the WHOLE WAY!!! :) But the details of distance and time will best be answered by John and Dale as well as some of the others I'm sure.

PS: DAVE AND ROD: I'll be sending you proper hi-res logos this weekend!

Jonathan

On March 3, 2012 at 10:55PM, Mike Watson wrote:

Jonathan,

Copy me as well on the logos. I'll fit it on the maps if I can. It looks like I'll have about 50 pages at a 1"=200' scale. This may be a more reliable method of accessing aerial photography. But, good be a little more cumbersome.

What brand GPS's will you all have? I'm hoping to provide reference gps points that you can load on to your gps units via a gpx file. This will allow you to have a better idea of where you are at with out a wireless network.

I really what to get this done before Ray's recon trip. It could be useful to know the 2010 photography actual shows current open channels. But, keep in mind in some locations it is not real clear where a channel might be. Would suggest you all look at the Bing aerial photography in detail. I've picked a route. But, you may see differently

Mike.

On March 3, 2012 at 11:23PM, Dale Sanders wrote:

Rod,

Bing and Google Earth maps will give you some clue. However, one can't mentally picture the dense swamp in some of the most God Forsaken terrane ever. Some places the river actually disappears (spreads out in a vast swamps). Now, having said that, I am 76 and feel I can make it. The first 18 miles will be the most difficult; shouldn't have too many problems after that.

I personally believe it will take us two and half to three days to reach New

Hwy. 72. (First 18 miles). So far, no one has ever paddled/ portaged craft the first 5 or six miles, from source to Blackjack road. Only two people we can find, John Henry Hyde and Bill Lawrence (now deceased) has paddled from Blackjack Road (about five miles down from source) to the Mississippi; John did it in "sections" over several years and Bill did it in one multi day "through" paddle). Two other people have gotten credit for paddling the full length - BUT, I have it in writing first hand, from Gary Bridgman himself, that he and Bill FitzGerald hiked the first 18 miles from Bakers Pond. There are some first here we can accomplish: First to "Through paddle" from source to Mississippi; First SUP paddlers ever to paddle the length of the Wolf; First to paddle/drag/portage craft the first five miles. We could hike like Geary and Bill did but I personally vote we leave our camping gear at the first night camp site and pull empty boats/ boards through. We might be able to do the same the second night and after that there shouldn't be any problems paddling with all our gear in boats/ on the boards.Now, to aide in navigation, in-other-words, making it possible to find the "Channels" Mike Watson has created, for us special maps, with GS coordinates, tracking the first 18 miles. We will print them out (20 or so sheets 8.5 X11), on water proof paper). With all the GPS's, smart phones and computers to show where we are we should be able to clearly see and find where channels there are. However Rod, even with the maps, this will not be a picnic.

Lets you and I, just the two of us, and possibly Mike Watson, the map maker, drive out (80 miles) and take a look around shortly after you arrive. We will then be better able to approach Jonathan with firm recommendations whether we hike, paddle or drag boats the first 18 miles. Lets do this on the QT, at least at this point.

On a related matter, I am planning on having lunch with "John Henry" Hyde early in week. He is very sensitive over his name and couple other quarks but I now recognize that and there so far has been no lasting damage. You know how Emails are - they sometimes get me in trouble, I believe, partially because I worked with the US Military for 37 years. You know how they get right to the point, which sometimes offends certain *personalities. Anyway, JHH is only paddling with us the first five miles, or possibly first 18 miles at max. (He told me several times, all he wants to do is paddle source to Blackjack, and possibly New Hwy. 72, so he can say he's paddled the full length of the Wolf). I just have to be more careful what I say to him in Emails. You might wish to share this Email with Dave etc., but it would likely be counter productive if JHH got hold of it. At this point, I believe he would be offended with me for even using his initials vice just "John Henry" - He's an artist several people have informed me. Mow, because of John's experience on the upper Wolf, he might actually answer your questions better than I. Will be interesting to see how he addresses the questions.*

Please let me know if you wish more details into the upper Wolf conditions. If you have a cell phone that does not cost to call USA - call- **********. Please give my phone number to your Dad and Sister.

Very Respectfully, Dale

On March 4, 2012 at 12:35AM, Rod Wellington wrote:

Dale,

Thanks for the heap of info! Dave C and I were chatting this morning about the distances on foot. Good to know that the 18 miles will be spilt over two or three days. We were wndering if that distance would be tackled in a day. Glad to hear it won't. Just so you know: since you're so keen to do it at age 76, the oldest among us will lead the team during the first two days! *Smile*

Happy to hear that you and John Henry are meeting for lunch. Hopefully you can both come away with smiles and handshakes. John's knowledge is certainly appreciated by this Canadian river runner. He has offered plenty of logistical assistance so far. I want you to tell him in person that I truly appreciate his input. It would indeed be a shame for him to step aside and not be involved in the SUPAS expedition. He is a good man in my eyes, regards of his personality. As you well know, it is exactly that uniqueness that makes us all special to each other. His passion for the Wolf shines bright in both his voice and his words. I honour that passion deeply. We may all be cut from the same cloth but the way we were stitched up certainly differs. You are all great people in Memphis, and that is the main reason why I wish to return to your town my friend.

Cheers,

Rod

On March 4, 2012 at 9:06PM, Dale Sanders wrote:

Quickie on the maps. Mike Watson and I drove out today to make sure the map coordinates matched to actual. Right on within couple feet. Wow! Were we happy. We will have 40 pages, in great resolution, of upper Wolf, from half mile below Bakers Pond to Michigan City —- Don't need maps after that. Total of 60 mega bites, jpg files. Do either you or Dave have a GPS on which we can load the maps? Also, I would like to have each printed (40 pages) on water proof paper but cost about $80.00. Could we possibly find a sponsor on your end. I will ask Jonathan also. Mike, earlier today, you mentioned getting others to look them over. Could we Email these maps possibly, to Rod and Dave etc.?

Jonathan has still not answered my question couple weeks back. "You need to assign a trip leader". He mentioned me lightly, nothing confirmed. Someone needs to start taking the lead and putting things together, even if it's John Hyde. Perhaps you know more than I. Hope so ☺.

I lost track of where Dave is right now. If he can receive Emails, please feel free to forward this on to him

On March 4, 2012 at 9:42PM, Mike Watson wrote:

I think I need to break the map sets down to five pages a set and send 8 emails. Each would be 7 or 8 Meg's. The imagery on the maps is from bing maps and dated as 2010. Bing had the best photography of the upper wolf.

The areas above and below Black Jack road are two areas that I believe will be the most challenging along with the first two. I'm going to review the areas again tomorrow evening. In all you have 22 or 23 miles to get to Michigan city. If I was to make a guess, I would say about a third of it will be walking.

GPS units: If you have garmin units, I can provided gpx files that contain tenth of mile markers and a proposed path. These features will be plotted on the maps and can be used as reference points on the gps units. Gpx files are interchangeable files and other units should be able to read them. I just don't have any direct experience.

Mike

On March 5, 2012 at 9:14AM, John Henry wrote:

I am proud to say that yesterday - I reconnected with my friend David Houston (an Ashland Mississippi local) who agreed to help BOTH trips by: either allowing us to camp on his property (in case that's as far as we make it on April 8th, OR - agreed to haul all our camping gear on his trailer, behind his 4-wheeler to "Number 5" (as the locals call it)... Number 5 is the pipeline which meets the Wolf River @ the Holly Springs National Forest, just upstream of Grogg Creek. Its legal to camp there, and if memory serves me correctly - has grass growing on it. So - now, BOTH of the necessary Headwaters campsites are secured! David said he would help us out for free, but - since we will need him to bring us our gear, and then retrieve it again th next morning - i feel it only fair that we pay him something for his trouble. I am sure that we can all chip in and make this worth his while! Two camp sites are necessary, as it was impossible to make New 72 from camp 1 in a day's time. :-) That's the hardest part of logistics for this brutal area, and how we get our gear from Number 5 back down to New 72 is the ONLY facet of this "Gem" of a trip which we lack - as far as logistics go. We will necessarily need someone to volunteer to retrieve my truck (loaded with all our gear, and hopefully more food, and icy-cold beverages) from David's, and meet the whole party @ New 72. By the time we reach that point - we will ALL need several beers to celebrate our major accomplishment thus far! :-)

I must remind ALL of y'all that the Headwaters area of the Wolf River is the MOST inhospitable, LEAST- User-Friendly place I have ever been, so - be mentally prepared for 3 days of snake-infested, chigger-infested, BRUTAL HELL - sprinkled with an occasional spot of beauty in-between the beaver-dams, log-jams, and tree-tops :-) I welcome this opportunity for ALL of you to prove to yourselves exactly what you are made of! :-) This will be an exercise in "Group Dynamics" in the highest degree! :-)

I went for a little leisurely paddle yesterday, from New 72 upstream - as far as the Old Highway 5 Bridge, and it flashed me back to the last time (7years ago) that I passed through that Godforsaken bottom... 7 hours, and my WHOLE body aches! :-)

Well - to quote Forrest Gump, " I guess that's all I have got to say about THAT!"

Dale Sanders replied:

Sounds like a leaders response to me. Well said. ☺

Rachel Sumner replied:

Goodness, thanks for your continued scouting trips, John! This is amazing work and the needed grit and grind to pull this event off. Guys - all of you - I know this is a great adventure for all of you (be mentally prepared for 3 days of snake-infested, chigger-infested, BRUTAL HELL - sprinkled with an occasional spot of beauty in-between the beaver-dams, log-jams, and tree-tops :-)), but the fact that you all are on board to intertwine this to be an event for awareness and support for what we're doing with OBS - THANK YOU.

Dale Sanders

Dave Cornthwaite

John Henry

Jonathan Brown

John Henry standing in the Wolf River near Moscow, Tennessee.

You guys are amazing!

On March 5, 2012 at 12:21PM, Keith Kirkland wrote:

Mike,

It sure is great having a volunteer for the WRC who is also a map maker!

John, congrats – you've found the true source of the Wolf.

I'm not in a hurry to get the word out, because Baker's Pond is such a great hike, but apparently your right!

Thanks guys for all the time and effort you put into the Wolf River Conservancy!

Keith Kirkland

On March 6, 2012 at 1:48AM, Dave Cornthwaite wrote:

Hi guys,

I need an address to have the main Aquapac order sent to...

Cheers

Dave

On March 6, 2012 at 9:45AM, Dale Sanders wrote:

Suggest my place. You and Rod will be staying here; (1) My Son's shuttling us out, early AM on 7th; (2) My truck is 4w/drive, 3/4 ton, with boat racks and canoe/SUP trailer; (3) All team members can bring their craft over Thursday or Friday (5th or 6th) load crafts & supplies (The John Ruskey boards will be here already) (4) We can hold a last minute Team meeting at my place Friday Evening. Info'ed Rod here for he and I have discussed the need for an all team member meeting prior to paddling): (5) Every team members would have the option to camp out in my basement if desired), so we can get a coordinated early AM start. (6) My wife Meriam is on-board with this. —- Ha!- One of the more important facts - ☺ . (7). We will need a Base Camp like area Memphis, so to speak - - My house can be that.

Meanwhile, in London, England, ol' Corn was boarding a plane en route to Mexico. Why Mexico? Glad you asked! He was about to begin expedition #5 in his Expedition1000 project (25 non-motorized journeys, each one at least a thousand miles long—or greater). #5 was a 3156-mile sailboat trek from Mexico to Hawaii aboard the Pangea Explorer. The adventure doubled as an interactive workshop/TEDx on the water, complete with daily presentations and a crew of adventurous souls who had signed up for the two-week experience. Once in Hawaii, the group would go their separate ways—most of them returning to the UK. Corn would board a plane to Memphis, paddle the length of the Wolf River with the SUPAS team, and then hop on the Bikecar and pedal to Miami. Whew! That's one busy Brit!

I managed to catch up with him via email in Mexico. He sounded overjoyed!

Hey Dave,

Did you have a date (or Memphis date) in mind for leaving Memphis heading south with the Bikecar? I know there are some events planned for the week following the Wolf paddle. (I'm planning to stick around and be involved in those in some way.) I'm trying to get an idea when I will be returning to Ontario during April.

Bikecar update: I received one offer of $1530, which I politely declined. Holding firm on $400 or lower. Pushing the charity aspect, hoping someone will take it for free or at a discounted rate. If no reasonable offers by *mid-Sunday I will begin looking at other shipping options. Hoping to have the bike Memphis-bound before you set sail.*

Enjoy the sand and sun!

Cheers,

Rod

Dave Cornthwaite replied:

Hey dude,

You'd love it down here, have you ever been to southern Mexico! Sea is gorgeous! And the boat is yummy!

Leave Memphis towards the end of the week starting the 16th, good to be around for any events happening after the paddle, although I'd imagine these would be pretty soon after we get back to Memphis. 18th/19th/20th – enough time to get prepared without it being TOO long.

Cheers for chasing shipping mate, really looking forward to April!

Now though, head back on the sail…

DC

In the days that followed, I sent out bid requests to dozens of shipping companies in hopes that one would ship the Bikecar for free in exchange for a heap of free publicity via social media. None stepped forward. In fact, almost all the bids were declined or unanswered. With the lowest bid sitting at $1530, I started to look into other shipping options.

Meanwhile, in Memphis, SUPAS preparations rolled onward.

On March 7, 2012 at 11:51AM, Rod Wellington wrote:

JB,

Will Luke Short be caneoing, kayaking or SUPing?

Cheers,

Rod

JB replied:

He will be SUPing. That makes four crazy people in the middle of no where standing on dang surfboards with dang paddles!

JB

On March 7, 2012 at 4:10PM, Mike Watson wrote:

Rod,

Attached is a preliminary Upper Wolf gpx file. Your supposed to be able load this file into any GPS unit, I believe. But, I only have experience with the garmins.

Typically garmin delivers an application called MapSource with the GPS unit. Open the application and go to file open on the menu. At the bottom of the Dialog box that opens is a pull down labeled "Files of type:". If this pull down is set to anything other than "All Files (*.*)" or "GPS eXchange Format (*.gpx)", select one of the two file types. Then browse to the folder in which you downloaded the file; select and open the UpperWolf.gpx file. Transferring the data to and from the unit is the same as normal.

Let me know if you are able to load it on a gps. I have an old eTrex Vista

HCx with four gig micro card and it loaded without a problem. If you are able to use this format, I will finish up, adding the lower wolf and the alternate routes for the upper wolf. This file has about 1000 points in it and if you have an older unit with out any extra memory this may be at the limits of what can be loaded. It wouldn't hurt if you would send me the make and model of your gps units along with a note about any additional memory that you may have purchased.

This has been a fun little project and have written some code that may be of some help to my co-workers as well. It would have been nice to have had this last year during the floods.

Thanks,

Mike

On March 7, 2012 at 3:24PM, Rod Wellington wrote:

Greetings,

The link to the Zero Emissions Expeditions SUPAS page has been updated: www.zeroemissionsexpeditions.com/supas-2012-2 (N.B. This link is now defunct.)

Cheers,

Rod

Rachel Sumner replied:

This is ridiculously encouraging! Thanks so much for getting the word out there, too!! That's incredible. The ball is rolling and gathering so much steam here it makes it unbelievable. It should be an AMAZING event!

On March 8, 2012 at 10:17AM, Dale Sanders wrote:

Dave/Rod,

To catch Dave, while in civilization, thought I would pass on some update info.

Everything is cool here. John, Jonathan and Rachel attended Canoe Club, (BCCC) meeting last night. I did the introductions and properly promoted your Web sites, in the process. All SUPAS folks left smiling. Club members very enthusiastic. Was 70 or 80 folks (paddlers) in attendance.

Jonathan will be sending out an update Email this weekend. SUPAS Web site is up as of yesterday. Maybe more sponsors will now commit.

Ray Graham, John and Richard will canoeing scouting trip next week. From Goose Creek Bridge to New Hwy. 72. Will certainly know more after that.

Wan't to personally send thanks, appreciation for all the on-line support you folks have already given. Now, if I could only say the same for OBS. They still, as of couple minutes ago, are not recognizing sponsorship, other than the link under "Recent Post". Hopefully my concerns will be OBE soon. Sure wish they would post some Home Page reference to SUPAS under "Upcoming Events". So far I have not been successful.

Richard Sojourner and I were to Scout today, bad weather - postponed to this coming Saturday.

Mike Watson's maps are a hit - hone run for us. Great detail. We shouldn't get lost, which likely accounts, somewhat, for the historically long upper swamp travel times.

John Ruskey is cool. He has the letter requested. A great guy. I really wish he could have paddled with us.

Again, thank you very very much for supporting out little local efforts here. We all appreciate you guys mucho.

On March 8, 2012 at 7:33PM, Mike Watson wrote:

I completed the gpx file. General river route should show up in red and alternate routes in green. Mile markers for the first 75 miles of the river and tenth of a mile markers on the upper Wolf.

On March 8, 2012 at 10:41PM, Robinson Littrell wrote:

Hey guys, I understand Dale is working on a rough timeline of when/where we'll be throughout the week. Dale, I'm so pumped about this! We can most certainly utilize this to our advantage! As soon as you get this completed, could you email it to us and let's get some extra eyes on it and collaborate? Thanks you guys for the work and effort you've been putting into this event!

—

Robinson Littrell
On-Campus MOVE Director
Operation Broken Silence

Dale Sanders replied (March 9, 2012 at 11:16AM):

Am already working on it. Will make final adjustments when Ray and folks finish scouting trip next week.

Anyone have recommendations, please Email me with suggestions. I have spent considerable time studying sections and find, reaching St. Georges HS by Friday 1400 -1430 could be an issue, depending on several factors. Hopefully we can reach New Hwy. 72, by mid day on Monday; but, just in case not, I have calculated, reaching New Hwy. 72 as taking full three days. Please check my math as well:

<u>Tuesday:</u> Mile Marker 83* to Mile Marker 65 (Ghost River Lunch Stop Area) - 18 Miles
<u>Wednesday:</u> Mile Marker 65 to Mile Marker 48 (Between Moscow and Rossville) - 17 Miles
<u>Thursday:</u> Mile Marker 48 to Mile Marker 31 (Just south of Collierville Arlington Road) - 17 Miles
<u>Friday:</u> Mile Marker 31 to Mile Marker 24.5 (St Georges High School, and on to Mile Marker 17 (Between German Town Pky and Walnut Grove Road) - 14 Miles
<u>Saturday:</u> Mile Marker 17 to Mississippi - 17 Miles
*New Hwy. 72.

Times/ millage goals are strictly estimates, depending on weather, river level, paddler physical conditions etc. By far, the most difficult day (on this schedule) will be Tuesday. We truly need to paddle beyond New Hwy. 72 on Monday, if at all possible.

Have copied Dave & Rod for impute/ suggestions. Also added Luke, for you to review.

Mike, thank you so much for the maps. Without them, I could not have put this schedule together. I am certainly most impressed with the quality, detail and data. Wish you could paddle with us.

When finalized, On Saturday 17th I will smooth up the schedule and put on spread sheet. Hopefully you will be able to add graphics to make it look better. You still didn't tell me what type craft you will be paddling from Michigan City and how far you will be paddling with us? Please let me know. I need to factor in every variable I can come up with.

VR, Dale

On March 9, 2012 at 1:31PM, John Henry wrote:

I am still willing to spearhead/guide the high schoolers race section @ St

Georges -if y'all cannot be there on time... My boat/s (3-man, and 1-man canoed) and I can handle that responsibility, provided we get a couple of WRC/BCCC volunteers to run sweep/support for those who flip. I am sure some of those kids will get wet. I am trying (through one of his top employees) to get the owner of the Daily News to jump in on this section as well.

As to the scouting Recon trip - I have closely observed the Headwaters area - as per the Bing Map APP I downloaded last night on my IPhone, and it will most certainly be one Helluva adventure! I feel kinda like a modern day "Lewis N Clark" type... The "Corps of Discovery" trip on the Wolf will be MOST memorable! :-)

Thanks for everything - John Henry

On March 9, 2012 at 2:11PM, John Henry wrote:

The Lagrange Mayor just told me to put him down as a "tentative" SUPAS participant, from the launch @ Lagrange at least as far down as the Rod n Gun club! :-) He said that if his health doesn't allow his participation - he can arrange for someone to represent his office, and his fair city as a participant. Now - we can use his commitment to leverage/pressure ALL the mayors along the Wolf River to join in on this EPIC adventure!

I thought that mayoral participation would persuade the Press to be more involved, and am Proud the Bill has jumped "on board"! :-)

If we pull ALL our individual "strings" - maybe we can get ALL the Mayors from Lagrange, Moscow, Rossville, Piperton, Germantown, Shelby County, and lastly Mayor Wharton, from Memphis involved! The more Mayors - the MORE press coverage for OBS/SUPAS, WRC, BCCC, AND the WOLF RIVER itself (what, Aside from OBS's mission, I hope, this trip is ALL about)! :-)

Sincerely, John Henry Hyde

Rachel Sumner replied:

John Henry!

You're continuously so impressive with wharf you get put together!

This is exciting!

Dale Sanders replied:

Once again John's enthusiasm is greatly appreciated. Having the Mayors involved could prove highly productive. Richard, Mike and I met one of the "Land Owners" we crossed yesterday. David Allen. He was very cooperative and showed interest and lives in LaGrange, TN. He might be a good person, to suggest to the LaGrange Mayor, to paddle a section with us, representing LaGrange.

SUPAS – Day 3 – April 9, 2012 – Easter Monday

    John Henry withdrew his wooden canoe paddle from the river and laid it dripping atop one of his craft's thwarts. He heaved a heavy sigh, braced his boots against the bottom of the hull, and stood straight up in one sinuous motion. Suddenly, he was the tallest thing in the swamp. He lifted both hands to his brow and shielded his eyes from the glaring sun as he scanned the slough for a navigable channel.
    "Movin' water," he'd said earlier in the day while perched in the same statuesque stance. "I'm lookin' for movin' water!"
    Somehow, he'd found some then and I had no doubt he'd find some this time as well.
    "John's good at locating the current," I thought as I sat slumped and waiting on my lime green paddleboard. "But he's not too good at keeping it."
    I looked over at Dale. His canoe was empty except for an expensive camera, a snack bag, and a jug of water. His camping gear, like mine, was locked up in the bed of John Henry's truck, five miles downstream

at the Highway 72 bridge. If we didn't reach the bridge by sundown, we'd be sleeping in our boats in the middle of a snake-infested swamp.

Dale wasn't standing and scouting. He was twitching in his seat like an impatient toddler in a shopping cart. He wiggled and wriggled and jiggled and squirmed. He fretted and fussed and tossed and turned. I thought he would leap clean free of his boat, and dance 'cross the lily pads like a gravity-less goat. But stay in his seat he did against choice, as he pursed his lips purple and silenced his voice. His anger was evident, his confidence irked. Beneath his white hair, insanity lurked. I hoped not to see it, hoped not he'd show, how volatile and malicious or low he could go. He'd voiced it before, in spurts on Day 2, when John Henry was absent from this hullabaloo. But now John was near, within eyeshot, within ear, within uncomfortable range that filled Dale with a fear so irrational, so silly, so crazy, so *true*, that it spread wide 'cross his face like a cancerous hue, a blackened shade, a darkness made, from antiquated thoughts and redundant beliefs that festered in a mind saturated with grief.

Tis in my nature to counsel, to guide. Tis in my nature to cower, to hide. Tis in my nature to love, to support. Tis in my nature to abandon, to abort. Tis in my nature to leave the wolf pack, to stumble ever onward and never look back. But look back I do, sometimes I stare long, and wish for forgiveness, for loving, for friends staid and strong, who voice their displeasure with respect and esteem, who cast off their burdens and admit they were wrong. Surely, thinks I, there can be such apt closure, such wounds healed from truth, and truthful exposure of venomous behaviour that no longer favours the goodness of all who choose to partake in the complicated crossing of a shallow, dank lake that lies in a remote corner of an American state. In the middle of nowhere, nowhere exists. And with nowhere comes questions, and wringing of wrists. When wrists have been wrung and skin sloughed away, marrow, and morrow, will cease to decay. A new hope will spring from a font once kept shrouded, and a new path will arise, uncluttered, uncrowded. Surely, thinks I, the canoe-statue man with the ginger goatee, who stands here before us like a young cypress tree, will find moving water and set us all free.

"This way!" shouted John Henry, pointing off to his right. His gaze led to a tangled row of cypress trees that towered over the swamp. "Follow me!"

Dale looked unsure. Doubt furrowed his forehead and crinkled the dangling grey mustaches that doubled as eyebrows on his 76-year-old face. It was a face that had weathered more storms than the rest of our group had seen collectively. Based simply on longevity, Dale trumped us all when it came to experiences in life and on water. He was a river guide, used to leading groups on the Wolf. Now, he was being led, something he secretly hated. His anger had been brewing all day, all week, all month.

In the split second before John perched himself back on his canoe seat and began paddling toward the towering cypresses, I carefully watched Dale's wrinkled face for a tell-tale reaction. There it was: the headshake of disagreement, followed quickly by the scowl of impatience and the glare of resentment. He was a silent, seething mess. He was relinquishing power. He was following the leader. He was also following the only person among us who had paddled this section of the Wolf.

John guided us down a twisting corridor of cypress trees, their grey-brown trunks rooted well below the waterline. The river's rippled surface parted cleanly around each tree as the channel flowed under our boats. Fallen trees blocked the route on numerous occasions and we found ourselves atop them, hauling our paddleboards and canoes over the trees and back into the stream. It was tedious work.

Finally, we reached what appeared to be an impasse. The forest had pinched us tightly into a corner. John left his craft and scouted the area ahead for an opening. He returned minutes later with the news.

"It's a dead end. No way through," he said as he climbed in his canoe. "We have to go back."

Groans of displeasure were heard from the other paddlers, me included. It had taken almost 20 minutes to wind our way through the thick forest and over fallen trees. Now we would have to do it again, in the opposite direction. I was not impressed. In fact, I was livid. Our late morning departure had put us at risk of being stranded in the swamp after dark, and every delay we encountered upped that risk considerably. Our late start, had, in part, been John's fault. He and fellow paddlers, Chris Reyes and Phillip Beasley, had downed plenty of beers the night before and were to slow to rise the next morning. Dale, in particular, was incensed when the group lagged behind in breaking camp. Dale—ever the disciplined Navy man—was ready to leave at 7:30am. The others weren't as eager, much to Dale's displeasure.

The night before, John, Chris, Phillip, and Robinson Littrell (from Operation Broken Silence) had set up camp on a wide swath of cleared forest that marked the location of an underground pipeline. During the initial scouting trips, the SUPAS team determined that this swath would be the most logical place to overnight on Day 2 of the expedition. As John Henry found out, accessing the pipeline swath was impossible without a 4x4. If this location was to be the camp for Day 2, he would need to somehow get two canoes, a stand-up paddleboard, and a host of camping gear into a very remote area. Luckily, he befriended a nearby landowner named David Houston who offered to transport the watercraft and gear to the area in question using his truck and trailer. David's involvement solved a huge logistical puzzle in the planning of how to descend the upper Wolf. Trucking in the gear—and trucking out the camping gear the next morning—meant we could all proceed through a very difficult section of the river without the extra weight. Doing so also made it possible to leave early the next day, which was the plan. That is, until the booze got involved.

The pipeline swath (called "Pipeline #5" by the locals) had been cut through a hilly section of northern Mississippi, and here, where the pipeline crossed the Wolf, the terrain was hilly as well. The others had set up camp while Dale, Jonathan, and I plodded our way through thick forest and daunting swamp on Day 2. Their location choice was smart (right next to the rough 4x4 road that led to David Houston's property), but it also put us well above the river's edge. The walk from the river to the camp was uphill, and I didn't think too much about it when I landed there at the end of Day 2. It wasn't until the next morning that the significance of the climb became apparent.

Dale and I were standing at the water's edge waiting for the others to join us. It was Day 3. 9:00am. Dale had already been waiting for an hour. His patience had long departed.

"Rod, would you mind goin' up thare and see what's keepin' 'em?" asked Dale. "Tell 'em we're ready to go."

Dale's request wasn't a question. It was an order. At least that's how I interpreted it. (Orders, I might add, especially ones given by people who have already gotten under my skin, don't sit too well with me.) I wanted to tell him to go check for himself, but thought better of it. *Very* reluctantly, I trudged up the hill to check on the others. Dale, who now seemed like he wanted nothing to do with the rest of the group, stayed with the boats.

Four weeks prior, Dale and John Henry had been involved in an unspoken race to see who would finish the Wolf first. John had boasted loud about his quest to become the first to paddle the Wolf from source to mouth and he took every opportunity to goad Dale with the challenge. Dale accepted, of course, because, well, he's *Dale*. Dale is a stubborn old coot who won't back down from a fight if he thinks he can win it.

On March 10, Dale and two other SUPAS members (Richard Sojourner and Mike Watson) struck out from Baker's Pond (the river's alleged source), pulling and paddling their canoes for five hours before taking out at Chapman Road, about four miles downstream. The arduous trek had been dubbed a "scouting mission," as had the previous two reconnaissance trips. What made this one special for Dale was the fact that John Henry had yet to paddle this section of the upper Wolf. John only needed to paddle from Goose Nest Creek to Blackjack Road (a distance of about six miles) to claim the prize of being first. Dale needed to paddle considerably further—to the Highway 72 bridge, a distance of 20 miles. On March 10, Dale shortened the gap to 16 miles. If he could somehow paddle those 16 miles before John Henry did, *he* would claim the prize. John, however, had other plans.

John decided to one-up Dale by paddling from Goose Nest Creek to Blackjack Road on March 11. The fact that he wanted to start from

Goose Nest Creek showed that John was adamant about his claim that Goose Nest Lake was the river's true source. He now completely dismissed the Wolf River Conservancy's claim that Baker's Pond was the source. Dale, of course, stood by WRC's claim. Despite objections from John, Dale planned to start the SUPAS expedition from Baker's Pond. John, who had no desire to paddle the river end to end in one trip, chose to begin his March 11 trek from Goose Nest Creek, "the true source of the Wolf River," he claimed. By day's end on March 11, John planned to be the first to descend the whole river. He didn't care if he did it sections. He only cared that he *did* it.

Even if John did beat him to the prize, Dale could take solace in the fact that he would, by the time SUPAS ended, become the first person to canoe the Wolf from source to mouth in one continuous trip.

"John might get the paddling 'end to end' title," said Dale to me before SUPAS' launch on April 7, "but I will get the ultimate prize for my thru-paddle."

Rivalries, once born, never seem to cease.

And so, on April 9 (Easter Monday, no less. In the Bible Belt, no less.), I was stuck on a sloped swath of cleared forest with one slow-moving SUPAS teammate, two slow-moving guest paddlers, and one irately impatient, god-fearing, ex-Navy man who wanted nothing more than to go on alone.

John Henry had asked friend and filmmaker, Chris Reyes, to join the group to help document Day 3 of the expedition. Chris would accompany Phillip Beasley (an old friend of John's) in "Big Bertha," John's hefty tandem canoe. Along with myself, John, Dale, Robinson Littrell and Jonathan Brown, Chris and Phillip's inclusion would bring the group's total to seven for Day 3.

Chris was of Hawaiian/Asian descent. His personality was unimposing. He was mild, friendly, and respectful in speech and mannerisms. His long, straight, black hair—tied in a ponytail—lent an air of rebellion to his presence. In conversation around the campfire, I learned he came from a punk rock background, as had I. Arts, music, and divergent personal politics had long been part of his life. He ran an online concert calendar and entertainment guide called Live from

Memphis, and regularly collaborated on artistic events with the City of Memphis.

I also learned that Chris was a vegan. I, too, abstain from eating animal products, and was pleased to find kinship with someone among a group of meat eaters. Although I never asked him, I got the sense that Chris was not a church-goer. This, too, eased my mind and I felt less alone. I don't consider myself an atheist, because I do not believe in a "God" of any kind, and therefore don't use a title to describe myself. Placing myself in the American Bible Belt with a group of devout believers was a cringe-inducing experience. I spent decades of my life socializing with anarchists who rejected all forms of organized religion. Religion, to me, seems completely unnecessary. It is an outdated concept. I have never been a "spiritual" person. If I can't touch it or see it, it probably doesn't exist. I cannot, for any reason whatsoever, believe in a higher power. I find it very difficult to be around others who do not share my opinions. I respect people's beliefs and would never try to sway a person from their core beliefs. I refuse to shout from a soapbox. I am fine when I'm alone or with a group of like-minded people, but I get very uncomfortable and cagey around church-goers. I don't trust them. In some small way, I fear them. It was completely unnerving to immerse myself in a group of paddlers who believed in a supreme deity and regularly attended church. I felt trapped with people I didn't understand and could not relate to. I felt like an outsider, as I had all my life. I wanted to flee, to retreat to a religion-free haven where I could be alone with my thoughts. During this trip, my tent was that haven, and I enjoyed shutting myself away behind its nylon walls. The less interaction with others, the better. But on this night, with the seven of us circled around the campfire, I felt as if I could relate to at least *one* other person, and that helped alleviate the anxiety. I was grateful that Chris had chosen to join us. Meeting him was a highlight of my journey down the Wolf.

Phillip Beasley and John Henry have been friends since elementary school. The two of them, along with Chris, have paddled together on the Wolf many times. Phillip is a talented singer/songwriter/guitarist and a huge supporter of the Memphis music scene.

It's not uncommon to run into him at local clubs numerous times in a week. He also happens to be an amazing chef, as evidenced by the huge meal of roasted vegetables and thick steaks he prepared on the campfire at the Pipeline #5 camp. He didn't have to do that. He didn't have to purchase all that food, truck it in to a remote area, and cook it over an open flame. But he did. It seemed that his gift of food and friendly conversation was rooted in the tradition of "southern hospitality."

I'm sure there are theories as to why people in the South are incredibly generous, and I certainly won't try to list them here. But I will say that I think it has something to do with religion and a history of poverty. When you have little and your neighbour has less, you help that person, you give. And although I don't know much about the Bible's teachings, I do know that promoting selflessness and maintaining healthy social networks are two things that resonate within the Christian faith. In a broader sense, one could argue that these are good guiding principles for all people to promote, regardless of religious identity. As much as I harp on in this book about my disdain for bible thumpers and church-goers, I do have to admit that, for the most part, the people I've met in America's Bible Belt have been friendly, generous, and extremely helpful. It's almost unfair of me to criticize them. Of the six people I camped with at Pipeline #5, five of them have firm beliefs in a Christian god. The sixth—and I'm speculating here—may have spiritual beliefs outside of the Christian faith. I don't know. I didn't ask. Regardless, all of them, and every other member of the SUPAS planning team, were friendly and seemingly genuine to me, and for that, I'm grateful. To me, they were all River Angels.

There was, however, one River Angel that got under my skin.

Dale's impatience on the morning of Day 3 was almost unpalatable. In many ways, I wish he would've gone on ahead without us. He didn't like that the others had consumed alcohol the night before, and he *certainly* didn't like that they were taking their time breaking camp the following morning. Even though he felt that John, Chris, and Phillip's inclusion on Day 3 was slowing down the expedition—which it was—Dale resisted the strong urge to strike out on his own that morning. Like the rest of us, he waited until everyone was ready to leave, and then nattered in my ear about how we had wasted hours of precious daylight waiting for the others. For the most part, he kept his displeasure stowed, but it leaked out from time to time as the day progressed. He grimaced when we followed John Henry down dead-end shoots. He bickered with John about directions when the current stilled and the vast, lily-pad covered swamps seemed impenetrable. And he silently voiced his disapproval when he abandoned the group at day's end. In obvious disregard of those who were silently pissing him off, Dale went AWOL and paddled on ahead, reaching the Highway 72 bridge long before dark, and long before his teammates. He later confided that he'd fallen in the swamp while pulling his canoe over a half-submerged tree. Soaked and trembling, he decided to abandon the others and make his way to camp. The question that stuck in my mind was: would he do it again? Would he desert me and the others at a time when we desperately needed his expertise? Desertion denoted a coward, not a leader.

My trust in Dale was fading by the day, with the possibility of evaporating completely by the end of the expedition. In three short days, SUPAS had become a fucked up, guilt-ridden, religion-slimed, personality clash played out in a snake-infested swamp in fucking *Mississippi*. It was an ordeal with no escape. I hated myself for getting involved with a fucking group of church-goers (I mean, *really*, what was I thinking?), and, to add insult to injury, the fucking task of pulling and paddling a fucking plastic SUP through "the most BRUTAL bottomland in the Mid-South" was *way* fucking more than I signed up for. I'm a determined man. Once I start something, I finish it. I was committed to paddle all eight days and by _____ (insert your choice of deity), I was going to do it—even if I fucking hated every one of my SUPAS teammates in the process.

On March 11, 2012 at 3:21PM, Dale Sanders wrote:

In an attempt to further refining subject schedule Richard, Mike and I paddled/ portaged Bakers Pond to Goose Nest Road Bridge (GNRB) yesterday. We were actually able to paddle around half of the distance. Took us 5 hours and 41 minutes, of which we were on the move approximately for 4 and half hours. I now feel comfortable that we can reach GNRB in time to proceed on to our first night campsite.

Mike was able to track, not only where we had been but how close we tracked the proposed route. Very close, not once were we confused, unsure of where we were. GREAT MAPS !!! With Mike, John and Rays help, our preparation effort are impressive.

Believe John and Ray are out today surveying paddle options between between GNRB and Blackjack Road. When I get their impute, will include and pass on to you folks.

I will be "Out of Pocket" until Wednesday evening. Early Thursday AM through Friday, am planning on helping Ray with his exploration trip shuttle needs.

To see few photos/ video of yesterdays trip, log on Facebook.

On March 11, 2012 at 6:02PM, Dale Sanders wrote:

Jonathan,

Plans are progressing well, for our SUPAS event. You are a great Trip Leader; I am happy to be a Team Play with you as my leader. Some suggestions for your consideration are:

(1) You Retain Trip as Leader Status. I will do my part to support your decisions. Remember, how Louis and Clark lead their expedition.

(2) Believe you need to appoint Section Guides. My suggestions are:

• I volunteer to Guide - from Put-in to Goose Nest Road Bridge (GNRB)

- I will know more about that section than any other paddler.

• Suggest John be asked to Guide from GNRB to New Hwy. 72 - He will know more about those section than any other paddler.

• I would be willing to Guide the New Hwy. 72 to LaGrange Sections, if you desire.

• Richard has expressed and interest in Guiding either/or both LaGrange to Moscow or our final day paddle - Walnut Grove to Mississippi.

As you can see, my leadership style is to get people involved, where practical. Therefore, why not invite Luke to Guide a section(s)? Should make him immediately feel important, he likely will research/prepare for his section(s) enthusiastically. You can double up me anywhere you wish. Don't recommend appointing Dave or Rod as Guides, unless they are asked in advance. Also, If you know anyone else you would like to invite on-board SUPAS, with only seven of us —- why not go for more?

The above, my heart-felt impute, just "food-for-thought. Please feel free to phone, if you wish to discuss. We will be in Hot Springs tell Wednesday, but I will have cell phone with me..

Please excuse my frankness in this or any Email. I was trained the Military way and that's hard to get out on ones system, especially with 37 years leadership experience with the US Navy. You can rest assure, if I miss something, say something to offend or just simply make a mistake, all was unintentional. My heart is in the right place...

Jonathan, again my hats off to you. I could not be, more proud of anyone. The community is so fortunate to have young leaders such as your self. ☺

Very Respectfully, F. Dale Sanders

On March 11, 2012 at 6:10PM, Dale Sanders wrote:

Rod, I blind copied you and Email sent to Jonathan. Just so you know, Jonathan appears to be dragging his feet on team guidance. That partially might be my fault, cause of the problems with John earlier, that have now been all worked out. Also, I get a strong feeling you and I share the same leadership values and lifes principles. I can tell you truthfully - I am very pleased you joined the team. Looking forward to you and yours visit and putting you guys up here in our place.

It wasn't easy by any stretch of the imagination, but on March 11, 2012, John Henry successfully paddled from his alleged source of the Wolf River at Goose Nest Creek to Blackjack Road, thus becoming the first person to paddle the entire river from end to end (albeit in sections). The race between John Henry and Dale Sanders was over. Dale, however, could rest soundly knowing that in a few short weeks he would attempt the first "thru-paddle" descent of the Wolf. For now, he relegated himself to shuttle driver for an upcoming scouting mission on March 15 and 16. In the meantime, John Henry and his paddling partner, Ray Graham, chimed in with these email updates.

On March 12, 2012 at 10:13AM, Ray Graham wrote:

John Henry and I did from Goose Nest Creek to Black Jack Road. It took us about 8 hours of tough going. There were nice stretches of open water, but most of the time we were dragging our boats through stands of heavy brush. I suggest wearing waders, and you might want to think about skipping this section.

Ray Graham

On March 12, 2012 at 11:28AM, John Henry wrote:

Yesterday - Ray Graham and I put in on the uppermost point of Goose Nest Farm's Creek and drug/paddled our canoes down as far as Blackjack Road. We did this trip blind - ie: we used no maps and had no GPS capabilities. The App on my phone was non-functional, as the GPS worked, but without a cell signal - the maps wouldn't download. (somewhat frustrating! :-))

This trip had occasional Beauty, and open-water (easy paddling), BUT - for the most part, and ESPECIALLY from Chapman Road to Blackjack Road, it was a BRUTAL push through major stands of Buckbrush ! We saw no snakes, but It was too cold for them to be out, as the last few nights have had lows around 40F, and the sun stayed behind the clouds most of the day. It took appx 8.5 hours with one lunch stop, and two rest stops to take a leak and recuperate from the exhaustive labor of bushwacking.

Snakes will be the one BIG issue for SUPAS, as the stands of seemingly impenetrable Buckbrush are perfect places for breeding and concealment of Water Moccasins.

Each SUP team member needs a tow-rope on front of their paddleboard, and each paddler necessarily should have chest waders, as the mud is often 2-3 feet deep with a foot of water on top of that. There will be several MILES of this trip where the paddling of ANY watercraft is impossible (due to the buckbrush), and proper footwear is IMPERATIVE! Any ideas on a Bass Pro, Lacrosse, or Red-Head wader sponsor for the team? Waders are a MUST, at least as far down as #5 pipeline (camp 2), and generally cost between $100.00, and 250.00 -depending upon model and manufacturer!

I am Proud to be associated with the SUPAS epic Journey down the Wolf River -as I have lived my whole life seeking out the next day's adventure, and I must say that y'all are all in for a GRAND adventure. The Headwaters Area of the Wolf River will test your mental and physical resolve upon MANY different fronts! :-)

Mid-week, Ray,Richard,Mike, possibly others, and I will be canoeing on down to New Highway 72 in conclusion of our Recon expedition, and images, videos, and additional info on paddling conditions will be made available as soon as is possible.

Sincerely - John Henry Hyde (Advance Scout/Guide) :-)

Richard Sojourner replied:

John Henry,

It sounds like fun was had but not by ya'll. Your description of what you and Ray encountered yesterday sounds much more severe than what we encountered on the upper section Saturday. Just wondering, if there are points above where the buck brush thickets start where we could access dry ground and portage around the thickest sections, we did this several times on our outing. The maps may reveal such areas. I can't stress enough how helpful the maps and GPS were in charting our route, Mike's done a great job there.

Rest assured John, your knowledge, experience, and enthusiasm are a huge asset to this endeavor and are truly appreciated by all involved.

What are the plans for Wednesday - Friday, I phoned Ray this morning but had to leave a message?

Richard

Below is Ray Graham's account of his and John Henry's arduous journey from Goose Nest Creek to Blackjack Road on March 11, 2012.

3/11/12

In an effort to "paddle" every section of the Wolf River, we decided to put in as close to the true headwaters as possible. After much debate, altimeter readings and water flow estimation, a group of river guides decided that the true source of the Wolf River is east of Goose Nest Lake. Since this area is private, John Henry and I chose to start on Goose Nest Creek where it enters public land at Holly Springs National Forest. It was a mostly cloudy day with temps in the low 60s. The water in the creek was clear.

After entering the forest, we pulled our canoes about a hundred yards from the road through the woods to the creek. The red clay creek bank rose at least ten feet from the sandy creek bed. So, we moved along the lip of the bank until we found a small tributary that provided us access to the water. The creek was too shallow to float our boats with us in them, but the water could easily carry the boats as we held the towropes that were tied to the bow of our boats. John described it as walking a dog. At times, the boats would even move ahead of us like a dog. The creek was fairly straight with a white sand/red clay bottom. I assume that it had been dredged at some point in the past. We made note of our passing of the Baker's Pond Stream. This stream is about a third of the size of Goose Nest Creek. Thus giving legitimacy to the creek being the major source of the Wolf River. As we walked, smaller creeks began to enter the Goose Nest mostly from the left. This slowly increased the volume of water. Booker Creek also entered from the left and at this point we could paddle our canoes. We floated easily downstream and our only obstacles were a few logs over which we had to pull our boats. I perfected a trick that I had seen others do in canoes. I paddled hard towards a log and made it half way over then moved forward to change the center of gravity of the boat. I usually kneeled at the front of the canoe and paddled the rest of the way over the log. Sometimes I did have to get out. I slipped and fell at one of the log crossings, but luckily the water did not top my chest waders. John and I agreed that the best way to traverse this area is in single seat canoes while wearing chest waders. According to maps, the land here is public national forest, but we passed privately owned parcels with houses and deer stands.

We reached a point where to steep banked creek open into a broad plain where the water spread out. The channel split into multiple directions and disappeared into heavy brush. It was at this point that we realized that neither one of us could get a phone signal. We would not have access to online maps or GPS. We would have to go through blind.

We attempted to follow the channel with the most water. The ridges on either side of the valley were visible, but a definitive river channel was difficult to find. This caused us to zig zag back and forth through the brush. The water was too shallow to float and the mud was too deep to lead our boats. So, I ended up pushing my boat while putting my weight on the stern so I wouldn't sink above my knees in the mud. We trudged through different types of mud as we went. The worst was the dark green mud that rested below the layer of surface brown mud. When the green mud was combined with a web of grasses and dead water plants, walking became difficult. Some of the vegetation was actually pretty. For example, we entered an area of iron wood hummocks. The trees grew in tight clusters, and we pushed our boats under their spreading branches.

Gradually, the water became deeper and we were able to ride in our canoes again. The vegetation changed with the water. Dead annual plants now surrounded us. One particular species had small cones. There wasn't room to use our paddles. So, we had to pull ourselves through the brush by grabbing branches and pulling forward. Luckily, it was too cool for snakes.

Our channel grew in size as other rivulets joined it. Then we were in the old river channel. The government had cut a canal through the river bed 100 years ago. Over time the water has gradually returned to the old channel. We followed its serpentine curves realizing this was a beautiful area. There were rock ledges at one point, totally unheard of on the Wolf. The south bank of the old channel was covered in moss and holes riddeled the bank. One was even large enough for a human to enter. We speculated that maybe it was the liar of a bear. A short time late we reached the Chapman Bridge. We finished the first section in about 4 hours.

We decide to go ahead and try for Three Bridges since we were already here. The river channel continued for a time, but then we lost it. We were back in the brush. John termed it the "Hell Woods". At least at this point, we could float through it . The brush was too close together to paddle so we once again pulled ourselves through by grabbing the branches. We did this for well over a mile. At one point, we spotted a heron rookery. About a dozen large nests occupied a large tree. The great blue herons sat on the nests or perched nearby. We also went through a series of beaver ponds. The water would get deeper and still. We could hear running water getting closer. Once we got to the dam, we had to find the best runnel to the other side. We followed the small channel to the next pond. This pattern continued over and over. Sometimes, the beaver dam had a break and we followed the swift moving channel through the gap. A few of them looked like small waterfalls, and we called paddling them "running the rapids". We also began to see parrot feather floating on top of the water. Parrot feather is a floating water plant that looks like green feathers. I was behind John by about 40 yards, when I spotted an otter. He saw me about the same time and bounded away. At times, we pushed straight through walls of brush until we found a small channel to follow. Once again other channels came together until we were floating in the old river channel. We passed under a foot bridge, and I saw that the land had been cleared. This was the property of the land owner who given us permission to camp on his property. We saw his two Polaris boats, hunting blind and duck decoys. We decided to go ahead and try for the bridge that was about 30 minutes away. Unfortunately, when we left his property, we lost the main channel. We should have stayed to the right. Instead, we drifted to the left of the line of trees in the middle of the river. This brought us into the heaviest brush. As they say, " If you are dumb, you had better be tough". We toughed it out and continued straight through heavy vegetation. It started the sprinkle, the rain drops made a rattling sound as it hit the dry grass. I continued to follow John through the heaviest brush. He was talking the whole time. I don't know if he was talking to me, himself or some one else. I felt as if I were following an apparition. Because of fatigue, I was beginning to wonder if I was hallucinating. However, I continued forward. I could see the bridge before I could see open water. Instead of following the main channel to the right bridge, we came out of the brush right next to the middle bridge. This area is not easy to traverse. I have to warn people not to try to "float" this section or to cross private property without permission.

On March 14, 2012 at 12:06AM, Jonathan Brown wrote:

Aloha everyone!!!

It's been busy at SUPAS HQ. Sorry for any lack of communication to this point. I'm sure everyone was stoked to have a bit of a breather though. :)

RIVER ANGELS

FIRST THING:
Please keep in mind this email is being sent to many people who may not be interested in receiving a million responses. Be sure to reply only to those needed. MAHALO! :)

SPECIAL THANKS & RECOGNITION
- Joe Royer for providing some of the prize packages
- Mark Babb for providing rentals and gift certificates
- Chuck Thomas iii w/ AT&T for providing volunteers
- BCCC for providing volunteers and river expertise
- WRC for bringing in Dave and providing support and advice
- Dale Sanders, John Henry, Richard Sojourner, Ray Graham, & Mike Watson have been the most amazing group of men! They've been jumping around in the mud, playing with snakes, and drinking water in the marsh all for the sake of SUPAS.
- Richard Sojourner for taking on the amazing task of being the BCCC volunteer coordinator.
- Mike Watson for working overtime and made some most epic maps that will be our lifelines while int he most unforgiving areas.
- Dale Sanders for providing leadership and taking initiative in the details of this trip.
- John Henry for paving the way with the locals, establishing our most needed camp sites, and gaining favor with all sorts of people.
- OBS (Rachel & Robinson) for making this paddle more than a paddle but an event that will change lives!!!!
- Dave and Rod for giving their time, energy, and efforts to help us southern folk out. You have become true friends and part of our family!!!
- John Ruskey for generously allowing us the use of all 10 of his SUP YOLO Yaks!!!! Without him we wouldn't be able to "Stand Up" against slavery. :)
- So many have helped and played a role. I'm sure I've missed many. The good thing is, it's not over. I have time to say thank you again!!!!

STANDUPAGAINSTSLAVERY.COM
- It's finished and ready to be announced!!!!
- Please help us get the word out.

- If you see any problems please let me know.

SOCIAL NETWORK
Facebook:www.facebook.com/pages/Stand-Up-Against-Slavery/242282799194178
Twitter: www.twitter.com/#!/OBS_SUPAS
Flickr: www.flickr.com/photos/supas/
Youtube: www.youtube.com/user/obssupas
- Please start spreading this around. "Like" the FB page, "Follow" the Twitter page, and "Share" everything!
- The success of promoting SUPAS weighs heavily on how much it's noticed on the above social networks.
- If you know anyone in Memphis that has a large following on either social network please get them to help us speed the word.
- WRC, DAVE CORNTHWAITE (who's on a boat)**,** ROD WELLINGTON, ADAM HURST(Outdoors Inc.): Can you be sure to saturate your social network with SUPAS?

FLYERS & HANDOUTS
- The SUPAS flyers and handouts are printed and ready to go. Look for them at your local Starbucks, Coffee Shops, and local businesses.
- I've attached a copy of the flyer.
- If you need some flyers and/or handouts please email Rachel

LOCAL NEWS & PAPERS
- Jill Bucco with FM100 has agreed to help us out
- If you have any connections with our local media friends please contact Rachel.
- Supposedly Channel 5 has already mentioned our event on air! Not sure how but we're not complaining. :)

SPONSOR & FRIENDS UPDATE

The Current List:
Rainbow Sandals, YOLO SUP, Quapaw Canoe Company, AT&T, Ghost

RIVER ANGELS

River Outfitters, Outdoors Inc., SUP Memphis, Bluff City Canoe Club, Wolf River Conservancy, John Henry Photography, & Aquapac

Working on:
Republic Coffee, Ghost River Brewery

Needing Sponsors For:
Waders for the team, Hammocks, Prize packages, Snake Venom Kit (just in case) :)

- Rachel will have a more up-to-date list of all the above.

SUPAS TEAM & TRIP UPDATES
- We have clarified where our start will be on Sat, April 7th
- We secured camping spots for the first two nights
- We have gained tons of favor in every area we will be paddling
- The Lagrange Mayor will tentatively be joining us. We would like to invite each Mayor on the water. If you can help please contact Rachel ASAP.
- The U of M Hockey Team will be joining us for the final stretch!
- Check out http://standupagainstslavery.com/donate-participate/ to find out what races and events there will be and please spread the word.
- If anyone knows a way we can get a snake venom kit please let us know.
- There have been and will be more reckon trips to ensure we know what in the world we're getting ourselves into.

If I've missed anything… and I'm sure I did, please fill in the gaps. But keep in mind, include only those to whom it may concern and please do not "Reply All". I wanted to include others for the sake of recognition and wouldn't want a blessing to be a curse. :)

With much appreciation,

Jonathan Brown

Meanwhile, in Eugene, Oregon, the Bikecar was still resting comfortably in Paul Adkins' yard—exactly where it shouldn't be. I'd explored other shipping options, but none panned out. I re-listed the U-Ship advert and impatiently waited for a bidder to bite. On March 13, I finally got what I wanted. A solitary shipping dude going by the moniker "boothbaybouncer" offered a bid of $473. I shot a quick Facebook message to Corn, hoping to catch him before he sailed out of cell phone range. I got lucky.

Hey Sailor (and I mean that with the most hetero gusto that I can muster!),

Quick update before you set sail. We have a shipping bid on the Bikecar for $474 + $18 fee = $492. I'll see if I can get the carrier to lower his bid below $400 including the $18 fee. Of course, there is a possibility that others will bid on the bike before the bidding wraps up at 11:59pm on the 16th. I extended the bidding twice in order to generate more interest. Seeing that we have at least one bid right now, we can rest assured that the bike will be shipped next week. I will arrange the pick-up and drop-off details and arrange the carrier payment. Not sure if you'll be able to check email/FB/Twitter while aboard the boat, but I'll send an update regardless when the bike is shipped. Have a great trip!

Cheers, Rod

Dave Cornthwaite replied:

Rod, this is great. Thanks so much mate. Can't receive mails or facebook etc on board but will be back on land on the 30th/3ast. Super that this has come together, price is perfect, you're an officer and a gent!

With no lower offers by midnight on the 16th, I accepted Triple B's bid. Days later, "boothbaybouncer" still hadn't returned my messages. Sadly, Triple B had bounced me back to square one, and there was no way to let Corn in on the bad news. Maybe it was for the better. I mean, really, why spoil a guy's sailboat journey with

RIVER ANGELS

one email? I hoped to have brighter news for him by the time he reached Hawaii.

Back in Memphis, members of the SUPAS crew had just returned from a taxing, but informative, two-day trek on the Wolf. They were eager to share their findings.

On March 16, 2012 at 8:56PM, Dale Sanders wrote:

I have finalized the below "Daily Paddle Schedule". This milage/ camp sites scheduled starts at New Hwy 72. The first two nights, before that, we are camping in the swamps between the headwaters and New Hwy 72. The unknown is, where will we camp Monday evening? The below schedule is only practical, if we can reach New Hwy 72 by around 1400 Monday, which will allow us to proceed toward Michigan City prior to pitching our Monday night camp. Every mile we paddle down stream from the New Hwy 72 bridge toward Michigan city will reduce a mile off the Tuesday 18 mile figure. Hopefully we can reduce the 18 miles to no more than 15. (The remaining days, Wednesday - Saturday, will not be affected by this adjustment). Please feel free to post this on the web site:

Tuesday: *Mile Marker 83, New Hwy 72 to Mile Marker 65 (Ghost River Lunch Stop Area) - 18 Miles*
Wednesday: *Mile Marker 65 to Mile Marker 48 (Between Moscow and Rossville) - 17 Miles*
Thursday: *Mile Marker 48 to Mile Marker 31 (Just south of Collierville Arlington Road) - 17 Miles*
Friday: *Mile Marker 31 to Mile Marker 24.5 (St Georges High School, and on to*
Mile Marker 17 (Between German Town Pky and Walnut Grove Road) - 14 Miles
Saturday: *Mile Marker 17 to Mississippi - 17 Miles*

Times/ millage goals are strictly estimates, depending on weather, river level, paddler physical conditions etc. By far, the most difficult days will likely be Saturday - Tuesday. We truly need to paddle beyond New Hwy. 72 on Monday, if at all possible.

Respectfully, Dale

On March 16, 2012 at 9:12PM, Dale Sanders wrote:

Last two days very productive. However, I didn't paddle - just the old shuttle master. John, Ray, Richard and Mike paddled (two days) from Blackjack to New Hwy 72. I did hear that Mike's Maps were a blessing. What do you guys thank?

On March 16, 2012 at 10:59PM, John Henry wrote:

This evening - Ray Graham, Richard Sojourner, Mike Watson and I returned from launching @ Blackjack Road yesterday, and canoeing downstream to SUPAS camp 2 (#5 Pipeline) where we camped for last night. The campsite was a Great location, and will work well for SUPAS! Camp 1 is also centrally located, and VERY functional!

Mike has now plotted the best available course from the source/s of the Wolf River all the way to New Highway 72, and all his effort in mapping and plotting the course will prove to be invaluable, as well as save MUCH time in brutal bushwhacking effort!

My friend David Houston shuttled our camp gear with his 4-wheeler and trailer from his place to #5 last evening, and assured us that he would do the same for SUPAS.

Camps 1and 2 are both perfectly located, and now secured as overnight bivouac spots.

We awoke this morning, and paddled on downstream as far as New Highway 72... On the way - we passed what I have called ,since the first time I saw it 7 years ago, the MOST beautiful swamp in the Southeastern

USA! It is truly spectacular. Regrettably - it was raining and I made NO images of this Wonder of the Wolf River... As I cannot currently afford to replace my Nikon due to water damage.

I cannot stress enough that Water Moccasins are out now, and will increase in sheer numbers and aggressiveness as the days get longer and warmer!

Each team member must become familiar with 4 types of snakes, and be able to ID each quickly. Water Moccasin (both subspecies - YES, there are two types!), Banded Water snakes (as they closely mimic Water Moccasins in their colorations), and Copperheads (rare, but - not unknown to me in the Wolf River bottom)... Quick identification of these types of snake could save any one of us from being bitten, and subsequently requiring immediate evacuation!

Other than names of shuttle/support group members, and exact times and locations where they should meet SUPAS at each designated bridge, I am fairly sure that all other tactical logistical issues have been worked out. @ least 3-4 waterproof walkie talkies should necessarily be available for this EPIC adventure, as it may still be necessary to split up and find the best way through the often BRUTAL underbrush!

It has been my pleasure to serve as Advance Scout/Guide for y'all, and I look forward to further assisting this EPIC adventure as it unfolds! :-)

Sincerely -John Henry Hyde.

Dale Sanders replied:

Thank you John - very well put. Congratulations are in-order for you as well John. I do believe you are the first to have canoed from Goose Nest Creek, above the confluences of Bakers Pond Creak, to the Mighty Mississippi. Way to go. Folks, this is going to be an unusual, difficult and challenging adventure, to say the least.

Robinson Littrell replied:

John Henry, we could never have pulled gis thing off without you. Its been an honor, a privilege, and s blessing that you've been on our team. Keep up the I credible job.

Keith Kirkland replied:

Thanks John,

I look forward to paddling that section w/ you perhaps in May.

Thanks,

Keith Kirkland

Below is Ray Graham's account of the SUPAS scouting group's two-day trek from Blackjack Road to New Hwy. 72 on March 15 and 16, 2012.

3/15/12

Day 1 - Black Jack (Three Bridges) Road to Pipeline #5 camp

This is the final trip in the Wolf River head water series. Four of us (Richard Sojourner, Mike Watson, John Henry and Ray Graham) left from Black Jack Road, which the locals call Three Bridges Road. It is so named because the bridge over the Wolf is broken into three sections with partial causeways separating them. The name Black Jack appears on many maps, but the local do not acknowledge it. This is the longest bridge to bridge sections on the river, and we decided that it would take longer than a day. So, we planned to camp out at a midway point where gas pipe line number 5 crossed the river. Luckily, John Henry had made a contact with a local man who agreed to take most of our gear to pipe line 5 by four wheeler.

Dale dropped us off at the bridge. Here, the river appeared as a stereotypical swamp with long pools of murky water separated by clumps of thick grasses and shrubs. There wasn't a current, but we headed west since that was the downstream direction. The river was broken into many different channels, and we attempted to follow the flow of water. We gradually drifted to the right side of the river valley only to end up in a small pond next to a farm. It looked like a great place to fish after a hard days work in the fields. However to us, it was a dead end. We backtracked laterally across the marsh without finding a passage. John Henry wanted to follow the flow of water through a thick lattice work of bushes, reeds and tree limbs. We tried to move through it, but it was impossible. The bushes with the small cone like seed pods that we had encountered above Black Jack were present, but now they were joined by trees with limbs that shot out in all directions like spider webs. Mike and Richard referred to their paper maps and decided that it would be better to leave to river bed and pull the boats through a meadow to a possible channel on the left. John Henry and I chose to go forward to follow the water flow. John Henry gave Mike a walker talkie before we split up. The reception was terrible, but we could hear them to the left. It became apparent that they had found a channel while we were trapped in an impenetrable labyrinth of swamp flora. John walked to where the others were to see what they had found. I soon followed John's foot steps and noticed a water moccasin a couple of feet from his boot print. The snake was hard to spot and anyone who comes here must be aware that these poisonous snakes are present and active in this entire river bed.

We pulled our boats over a small piece of land to join the others in the left channel. On the way, I spotted a small salamander with a copper back. I stopped to take video. We followed the channel but had to deal with the repeated problem of having to cross a beaver dam and then finding a channel again. Sometimes crossing over the dams was easy and fun because of the small rapid created by the beavers. However, other times the dam broke the river into a hundred trickles and it was up to anyone's guess as to which stream would lead to a channel.

At last, we reached a point where the channels came together to form an actual river. John, who was the only one to have completed this section, exclaimed that he was relieved because the worst part was now behind us. We would only have to worry about log crossings and no more "buck brush". We were all happy to here this. When enduring hard travel, it is good to have a moral boost occasionally. A short time later we reached the confluence of Tubby Creek and the Wolf. It was near an old bridge crossing at Shelly Road. The old bridge supports still protruded from the water. Because of the heat, I decided to switch from waders to aqua socks and river shoes. The waders are good when it is cold. However, when the temperature rises, they are a hindrance. When the temps reach 80, waders are too hot wear. Any high boots may also be a hindrance. To complicate matters, snakes are also likely to be encountered at arm/face level. I'm not telling anyone not to wear high boots, but remember that you are not fully protected. You have to be very careful about where you put your hands even when you are exhausted.

For the remainder of the day, the only obstacles were log crossings and beaver dams. We passed a small bluff where John had said he had found a wild azalea bush seven years ago. We crossed the first two pipeline crossings and after six hours, reach our camp at pipeline five. The bank was low and swampy, and we found it necessary to walk up the grassy pipeline strip to the top of a hill where we could camp.

John went down stream to explore the rock bluff, and the rest of us stayed to make camp and wait on our supplies. While we waited a couple of young men rode up on four wheelers. We talked for a while. One of them said that he had grown up in Frayser, but his family had moved to escape the crime. About 30 minutes after they left, our supplies arrived. A local man with his nephew and wife drove up on a four wheeler pulling a trailer. Our campsite is only accessible by four wheeler and now canoe. We had thought that they could have dropped of John's truck, but that seems impossible. We talked with the family for some time. They told us stories of home made boat and snakes. They left about ten minutes before John returned.

That night, we cooked and looked at the stars. It seemed as if we could see every star in the sky. There was no light pollution at our camp, a rarity for all of us. We also saw a couple of satellites. It had been years since I had seen one.

3/16/12

Day 2 – Pipeline #5 camp to New Hwy 72

The next morning rain was threatening. I could get a signal from the top of the hill, and a weather page showed a storm possibly bearing down on us. Luckily, it looked as if it would just edge us. After packing, we headed downstream towards the first beaver dam. It was already in the 70s and the sky was full of rows of slate blue and light gray clouds, ominous but beautiful. The Wolf was now a real river, within a definite channel and flowing with strength. The beaver dams were also larger, but most of them had breaks with waterfalls some of which were almost two feet high. After two or three dams, we reached the rock bluff. It was not as high as I expected but still impressive. There was even a small cave at the base. I was the only one to get out and climb it. There was even a small island next to the bluff. It would not have been a remarkable landmark on most rivers. However on the Wolf, it is the only known rock bluff.

We crossed several logs and beaver dams. One of the dams split the river into dozens of different streams. They rejoined the river in an impressive cascade. A few miles later we passed an area that had been logged. It looked as over a hundred acres of trees had been cleared. Some of which were left lying in the mud waiting to be washed into the river by the next flood. We finally reach what John had described as the most beautiful swamp in the Southeast. I think he is right. We struggled to think of a name for it, but I couldn't think of one. The area reminded me of a hybrid of the Ghost River and the grasslands of the Bateman section. It was mostly open marsh but studded with tupelo and cypress trees. The river snaked through it. There was a nice contrast between the green meadow and the bands of gray and blue clouds. Barren trees stood out against the sky.

The Wolf reentered the forest, and we reached the old 72 bridge. The wooden structure was partially dilapidated, but the creosote preserved the timbers. Most of the bridge was still intact. We continued downstream for the next three miles to HWY 72. During this time, we had to portage several logs and a few beaver dams. We also passed a field with several geese. Unfortunately, I began to get sick towards the end. I had been fighting a stomach virus all week, and my body finally gave into it. The last three miles were very difficult. It would have taken me much longer to finish if it were not for the help of Richard, Mike and John Henry. I have to give a special thanks to Richard, Mike, John and Dale for making this trip possible.

The wildlife sighted on this trip include deer, beaver, geese, bats, a salamander, at least four water moccasins, non poisonous water snakes, king fishers, great blue herons, wrens, sparrows, turkey vultures and some other stuff I forgot about. On a final note, the sections from Baker's Pond/ Goose Nest Creek to HWY 72 are a wilderness area. There is a large variety of wildlife and very little trash, but it is harsh terrain. This part of the Wolf should only be traveled by experienced outdoors men/women. I do not care to return to most of this area. As they say in the exploring business, "My curiosity concerning this section of river has been satisfied". That being said, I would like to paddle from pipeline #5 to HWY 72 again. This section of river contains the highest beaver dam rapids, the rock bluff, a beautiful swamp and the old 72 bridge. I now consider it to be the best section of river. However, one needs a four wheeler to access the river at Pipeline #5.

On March 17, 2012 at 10:58PM, John Henry wrote:

If you see this view (a photo of a snake) - you are VERY close to being bitten! Their strike is lightning - fast! If you don't act right @ this point - your SUPAS adventure will soon be over! Know how to quickly ID all 4 types of snakes mentioned in last night's email before April 7th. This is a female Water Moccasin (Cottonmouth). She was less than 2 feet long, and I made her photo two days ago. She came @ my canoe from a long distance, and showed me her teeth when I did not retreat. This is not uncommon behavior this time of year. The other subspecies of the Water Moccasin has different colorations altogether.

The snake in this image and the non-poisonous banded water snake look almost identical, except head shape, and pupil shape. Water Moccasins won't retreat, as they have NO Natural Predators!

Banded Watersnakes WILL retreat, as they are often eaten by Water Moccasins!

Don't fear snakes - understand them!

On March 17, 2012 at 11:16PM, Mike Watson wrote:

Hopefully you all can open one of the two attached files. They are the actual route that we took through the upper Wolf from Bakers Pond to Hwy 72. The tracks document the start and end times of each day with the exception of Blackjack to pipeline 5. You can also get an idea how much time we spent resting or stuck. These locations are where the mph is at or near 0.

I forgot to turn the gps when we started at Blackjack. I did save a waypoint as we were leaving and the time was 14-MAR-12 11:06:05AM. We paddled 50 minutes prior to turning the gps on. I will digitize missing part of the route we took and replot this route on the maps that will we used for the April 7 trip.

Mike

On March 18, 2012 at 12:16PM, John Henry wrote:

Mike's maps will serve SUPAS very well! Especially from Goose Nest Creek down to Blackjack Road. Since the GPS wasn't turned for the "Gauntlet of HELL" on the Blackjack Run - that will still be the MOST BRUTAL hour/two hours of the whole week! SUPAS must pace themselves in the upper 3 legs and make sure to stay hydrated, so as to NOT be burned out on the last 4 days...

From new 72 down to Memphis will be a cakewalk compared to the Headwaters area, and taking time and enjoying and documenting the Headwaters Area - through as many photos and videos as possible, the unprecedented Beauty of the headwaters area is VERY important!

At least 4 waterproof walkie-talkies are VERY necessary for the Headwaters Area! We had to split up in the "Gauntlet of HELL" and the ones I brought were useless, as their range was insufficient!

Other than gear/shuttle specifics, and sponsorships - I feel all other logistical issues have been ironed out.

Ray Graham overheated in his Waders - and recent experience in those brutal conditions has proven that knee boots (I personally prefer Lacrosse Burly Classics - and have walked/waded thousands of miles in those very boots) and several pairs of quality lightweight wool socks for rotation will indeed work out better!

Richard Sojourner replied:

John Henry,

I agree about the boots, mine are mid calf height snug fitting Neoprene from the shoe portion up and although I went over the boot tops a number of times in mud they never tried to come off with wool socks and sock liners my feet didn't get/feel wet until I walked in mid thigh deep water several times.

Richard

Mike Watson replied:

I'm not sure getting wet will be a bad thing. I got hippo-thermic with the chest waders and ended up soaked with sweat. Snake guards would be the primary concern.

Mike

RIVER ANGELS

On March 18, 2012 at 10:24PM, John Henry wrote:

It took us (Ray Graham and I) 8- 8.5hours to go from Goose Nest Creek to Blackjack Rd. It will take SUPAS around 7-8 HARD hours to reach Camp 1 @ James Lowery's place. From Camp 1 to Blackjack =.5-.75 hour. From Blackjack Rd to Camp#2 (Pipeline#5)...will be around 7.5-8hours. This run is by far the WORST part of the river for around 2-3 hours, and holds an area which I have disaffectionately named "The Gauntlet of HELL"! The paddle from #5 to New 72 will take between 6-7 hours.

These estimates will vary depending upon many factors - the most of which is the amount of time you choose to spend documenting this journey... Personally - I feel that each team member team should take the precious time to bring back a clear record of this EPIC journey through the Headwaters Area of the Wolf River, as it is the least-known, and most Brutal region in the whole watershed, but - that's just MY opinion. Ya'll will each have different means/methods of doing so, but myself - on top of creating hundreds, perhaps thousands of still images, I have invited a professional videographer (Live From Memphis.com - Christopher Reyes) to join me upon my boat (the USS Phoenix) from pipeline #5 down and video the INCREDIBLE swamp appx 3.5-4 miles upstream of New 72. Extra logistics, I know, but - well worth the effort, as this one particular swamp is by far the most beautiful open, running swampland I have yet found upon this incredible river! Having canoed the whole river - i feel confident that you will all agree with me when you witness it firsthand! I look forward to sharing the Headwaters Area of the Wolf River with you, and hope that it is indeed a safe and enjoyable journey! :-)

Sincerely, John Henry Hyde

On March 18, 2012 at 11:47PM, Jonathan Brown wrote:

Hey all!!!

So I'm gonna throw a date and time out and let me know what you think asap please.

Meeting at Dale's house Wednesday March, 28th at 7pm.

If we meet at Dale's house it will be easier to Skype in Rod, Dave will be on a boat, we can pick up our new Aquapac bags, and we will all be familiar with the location of Dale's house for the actual event.

Adam, Rachel, Robo, Keith, Ray, Mike: I think having you guys there would be super important. PLease let me know if this will work.

Honestly it would take a miracle for all of us to meet but I'm hoping it can happen. This will be a very important meeting that everyone on the SUPAS team will need attend, with the exception of Luke who will be at school.

If there is anyone I forgot please let me know.

Jonathan

On March 19, 2012 at 1:26AM, Jonathan Brown wrote:

Aloha Y'all!!!

Now is the make or break time for spreading the word about SUPAS! If we get the word out effectively, we will raise an awareness concerning the event which will naturally bring peoples attention to the cause, the events, OBS, WRC, and more. This event has been designed in such a way that everyone involved wins!!!

This events full-on success is dependent on effectively raising awareness, participation, and simply having the SUPAS Team complete the journey. :)

With all that said, effectively getting the word out now is vital.

If you have any opportunity to connect with potential sponsors, local media, or need to communicate officially concerning SUPAS we would ask that you let Rachel Sumner know about this.

FLYERS & HANDOUTS:

Rachel and Robinson will be saturating our area with SUPAS flyers and handouts so be looking for them at you local coffee shops, eateries, and business places! If you need any to put up at high-traffic areas in Memphis please contact Rachel or you can download and print them by clicking the following links (N.B. Links removed).

SOCIAL NETWORK:

Using our social networks are highly instrumental in our efforts to spread the word. We ask, if you have Facebook and/or Twitter, would you please use them as marketing tools to help us spread the word?

I'm wondering if WRC and Outdoors Inc. could begin promoting this event through your networks? You may have already started but just in case.

Please post and tweet with wisdom and be sure to communicate which organization is hosting SUPAS (OBS) as well as what the purpose and cause are (raising money for an aftercare facility for human traffic victims through paddling the entire Wolf River and inviting others on this most epic journey).

Here's the social network info you will need:

SUPAS Online:
www.standupagainstslavery.com

SUPAS on Twitter:
Twitter Site: twitter.com/#!/OBS_SUPAS
Hash Tag: @OBS_SUPAS
note: the "@OBS_SUPAS" is called a "hash tag" in Twitter land. If you include this simple hash tag in any Facebook or blog post or Tweet, it will act as a link to the SUPAS Twitter page which will then motivate people to "follow".

SUPAS on Facebook:
Facebook Page: www.facebook.com/pages/Stand-Up-Against-Slavery/242282799194178

note: Use your Facebook page as a billboard for SUPAS. If you're on the SUPAS Advance Paddle Team and have photos, stories, and/or videos that will get others interested in this event please post them on the SUPAS FB page.

SUPAS on Flickr:
Flickr Stream: www.flickr.com/photos/supas/

SUPAS on Youtube:
Youtube Channel: www.youtube.com/user/obssupas

To show just how serious (and some might say *obsessive*) Dale Sanders became about paddling the Wolf River from end to end, he drafted the "Individual Paddle History - Achievement Record" form and invited paddlers who had completed sections of the river to submit the form to the Wolf River Conservancy. To me, the form served as a way to rank those paddlers who claimed "firsts" on the Wolf. Gary Bridgman and Bill FitzGerald travelled the length of the river in 1998, but they did not paddle the whole river—they apparently hiked the first 12 miles, and then paddled the remaining 90 miles to the Mississippi River confluence. Dale's form recognized their huge achievement (they are likely the first people to ever travel the length of the river under their own power), but it also minimized what they did in order to make Dale's great accomplishment appear even greater. Dale planned to paddle the river from end to end in one continuous trip (as part of SUPAS) and he planned to pull or paddle his boat every inch of the way. In doing so, he could claim to have paddled the entire river in one go, something that no one else had done. Fair enough. It's a big accomplishment and he should be recognized for it. *But*, I don't think it was right of him to devise a method of ranking people's accomplishments (based on their method of travel or whether they did the river in sections) in order to make himself look better. His form minimized John Henry's recent accomplishment. It divided camps yet again. Simply by listing a choice of two "sources," he recognized John, and insulted him at the same time.

Dale is an interesting character. He shows good intention, but I always feel there's an ulterior motive lurking beneath his good intention. Most of the time, that motive is to make himself look grand by influencing people to turn against other people. It's a power game. It's all about control. I've seen Dale do it a dozen times. I caught wind of his game early on and refused to play into it. I grew to distrust him and limited my exposure to him. Writing this book serves as a way of processing the disparaging feelings I have toward Dale Sanders. It helps me find a place in my heart where I can store the "idea" of him, where I compartmentalize him and access him when I want. That way, I don't have to deal with the animosity, resentment, and anger I feel towards him on a daily basis. Writing about it helps me purge those feelings, or at least diminish them. The resultant book becomes the storage device for those feelings. Most of the time, the book remains on the shelf and I do not interact with it. But sometimes, I take the book down and browse through the feelings. Then, I put it back on the shelf and move on. It's all about moving on.

I applied that same theory to my first book, *Part-Time Superheroes, Full-Time Friends*. Writing that book helped me put my friendship with Scott McFarlane into perspective. I had to find a place for Scott in my heart. It took 14 years and 288 pages to do it. That book remains on the shelf until such time when I want to revisit my feelings, feelings similar (and sometimes identical) to the anger and animosity I feel towards Dale. *Part-Time Superheroes, Full-Time Friends* provided me with the opportunity to move on. It's all about letting go and moving on.

With 18 days until the SUPAS launch, I booked a flight to Memphis. Hopefully, three weeks in Elvis' hometown would be enough time to sort out the Bikecar details prior to the Wolf River adventure and also give me a few days to pedal south with Corn before I headed back to Ontario. After all the time spent dealing with preparations, it would feel great to finally meet the SUPAS crew and wet a paddle. I was looking forward to it big time.

That same day, I got a hold of my old friend, Michele Glasnovic, and alerted to her to my plans. It had been seven months since my last visit to Memphis and I was eager to catch up.

On March 20, 2012 at 6:58PM, Rod Wellington wrote:

Hi Michele,

Hope all is well with you and Greg. Are you still living in Memphis? Moved to Portland or elsewhere?

I will be back in Memphis from April 2-25. I'm doing a slideshow presentation at the Bluff City Canoe Club on Wed. April 4 at 7pm. You and Greg are certainly invited. It would be great to see you both! (I will forward you the address where the presentation will be held.)

From April 7-14 I will be part of a team that will attempt to descend the Wolf River from its source in northern Mississippi to its confluence with the Mississippi River on stand-up paddleboards, something that has never been done before, or even attempted! Many of the same people that I kayaked with last August will be taking part in this expedition, including British adventurer Dave Cornthwaite, who we saw talk at Mud Island.

The expedition will also function as a fundraiser. Money raised from the SUPAS expedition (Stand Up Against Slavery) will help fund the implementation of an aftercare facility for survivors of human trafficking in Memphis, TN. Operation Broken Silence, a Memphis-based non-profit organization that strives to protect the innocent by confronting social injustices, speaking and acting against the growing threats of mass atrocities and modern slavery in our world today, will oversee the implementation of the aftercare facility. The Wolf River Conservancy is also involved with the expedition. Feel free to spread word of the SUPAS expedition/fundraiser to others.

For more information, visit the official SUPAS website www.standupagainstslavery.com and www.operationbrokensilence.org.

I'm also putting the finishing touches on a book called The Brothers. My trip down the Mississippi River in 2001 with my friend Scott forms the centerpiece of the book. You may recall reading some excerpts from the book last summer. I found a local artist to do the chapter page illustrations and the cover. They look fantastic! He's really good!

Before I proceed with the illustration for the Mississippi River/pontoon boat chapter (Minneapolis/St. Paul to the Gulf of Mexico), I need to ask permission to use your likeness. Because there were so many momentous events that happened during the river journey, it was very difficult to choose one image for the chapter page. But in re-reading the chapter I found that the time I spent in Memphis stood out the most. I really needed a break from Scott by the time we reached Memphis and was very thankful to have met you and spent time with you. This respite from Scott and the river figures well in the story and I think makes the best content for an illustration. So, what I'm thinking is to have a scene where you and I are walking away from Scott and the boat at Mud Island marina - you and I in the foreground, Scott is on the boat in the background. So, if you're comfortable with being part of this I would need to get you to describe yourself as you were in summer of 2001 or, if possible, send a picture of yourself from that time. Of course, I have my own description of you that I could relay to the illustrator. It may not be entirely accurate (it was 11 years ago!), but it would suffice. You may even be more comfortable with the fact that it is not entirely accurate! Anyway, let me know what you think.

Chat soon.

Cheers,

Rod

Michele Glasnovic replied:

Hi, it's great to hear from you!

We're in the final stages of preparing our house for the sale. Already have a realtor selected. We're just doing some fix ups so that the house sale goes smoothly and the home inspector has nothing to report but "great house, great condition!" If all goes well with the plumber who is coming out on Saturday to do the last major fix (we discovered a leak in the upstairs shower - yuck), it might be as early as Monday that we sign final paperwork and put a sign in the yard.

At this time, I can't make any promises to anything as to where we'll be or what we'll be doing - all of it hinges on how well this house sale goes. But we will be in Memphis in April, and if we're lucky it'll be in an RV while we do job searching on the west coast. We decided to, instead of going the traditional route with moving, to buy an RV and move in that across the country - and take a few weeks between ending work here and starting work somewhere else to enjoy the countryside. We're very excited about that. Any recommendations on places I shouldn't miss?

Your expedition along the Wolf River sounds wonderful, and congratulations on doing the slideshow event. If we can make it, we will. If not, I'd definitely like to catch up for dinner and hear all about what you're doing.

Nice job on the book. I don't know if I have any photos from me back then - don't know where they'd be (probably on a hard drive somewhere...) I'm happy that you're including me in the book, though. I can send you a more recent photo if that would work.

Excellent stuff about that fundraiser. I'm looking over the sites now - I hope it goes well with that. I'm glad people are taking note of this problem - it's something a lot of people just aren't aware of (or that it happens right here at home.)

Great to hear from you. Let me know how things are going when you're closer to ready to come out to Memphis!

On March 22, 2012 at 12:52AM, John Henry shared a link to

his newly created blog, *Where the Wild Things Are*. In the blog's initial post (dated March 21), John talks about his deep connection to the Wolf River and sets the record straight about becoming the first person to canoe the entire river.

The Wolf, The WHOLE Wolf, And Nothing But The Wolf - So Help Me GOD!

I am NEW to this whole blogging thing, so - please bear with me if I seem a bit Longwinded.

The first time I set out to see the Wolf River upstream of Memphis Tennessee - it was ONLY because I was in search of Native American artifacts... You see - I am really a Native American wrapped in white skin, and I have for most of my life sought out "The Things They Held Sacred". I come from the Sacred Smoky Mountains, land of the Cherokee, and my Great, Great, Great, Great Grandfather (Nathan Kirkland was his white name) was an elder of the Cherokee Nation and was present @ the signing of the treaty which forced the Cherokee upon the "Trail of Tears". He lied to the US government, and claimed a small percentage of white lineage, in order to remain in the "Sacred Mountains". He had, years before, followed the last herd of buffalo as they left the Cheoa Valley (site of present-day Knoxville TN). My affinity with the Wolf River is borne into me, and is indeed beyond my limited powers of comprehension...

On my maiden voyage to the Wolf River - I made it downstream from the boatramp @ Lagrange TN as far as the Eastern edge of the "Ghost River", and then decided to turn back. I had never before been on such a small, winding river in my 15'x 48" semi-V welded aluminum boat, with a 30 HP Mariner outboard engine, and the shallow water and numerous logs had me quite worried about destroying either my lower unit or my stainless prop!

The next time I came back to the Wolf was two years later, on a scouting mission to check on available boatramps for launching a canoe. My friend

Chuck and I went to the Batemen Road ramp just a few miles East of Moscow TN, and encountered a fellow who said that he had "Canoed the WHOLE Wolf River". He had, in fact, canoed from Michigan City Mississippi all the way to Memphis. As we drove away from that chance encounter - I said to Chuck, "Hell, man - there aint nothing special about That guy! If HE can canoe the WHOLE Wolf River, then - so can WE!".

I have, since that day, come to realize that to "Canoe the WHOLE Wolf River" means many different things to many different people!

In fact - the Wolf River starts near Ashland Mississippi, in a couple of different locations. The most-accepted Headwater of the Wolf River is undoubtedly Baker's Pond. It lies in the Holly Springs National Forest, on public land. If you ask the locals - they will tell you that Baker's Pond is in fact an area, and NOT just one body of water. The area is around 1/2 -2/3 mile long, and comprised of three separate, interconnected bodies of water. The opinions and observations of the "Locals" is always an invaluable tool when exploring what should aptly be known as a "Wilderness Area". WILD being the most important part of the word Wilderness.

In my quest to "Canoe the WHOLE Wolf River" - I came to believe that the TRUE Headwaters of this Primal, Brutal, Beautiful river originates not only @ Baker's Pond, but also and primarily in another location altogether... a lake which lies on private property which is named "Goose Nest Farm". The creek running out of and below this private lake has no official name, so - I have henceforth named it "Goose Nest Creek".

I alone made this determination about the TRUE Headwaters based solely upon what my eyes saw, what my mind thought, and upon Altimeter readings, and official National Forest Service maps. To question the authorities that be about the TRUE Headwaters of the Wolf River was somewhat of a touchy subject. Nobody had ever before challenged the popular designation by the early members of the Wolf River Conservancy (a non-profit organization whose original purpose was to protect the water and land along this grand river), but - I eventually had NO choice except to lay my theory out there so that John Q Public could either accept it as Fact, or - shoot it full of holes, and make me out to be just another fool with a paddle in his hand! Baker's Pond Creek and Goose Nest Creek meet on private property just a few yards below the southernmost boundary of one of many sections of National Forest in that area.

On Sunday, March 11th, 2012 - a friend (Tim Darkwing Parker) (N.B. Aka Ray Graham) and I hauled our separate one-man canoes onto Holly Springs National Forest in my truck, drug them around 120 yards through the dense forest, and descended into Goose Nest Creek in order to canoe for the FIRST time in recorded history - from the TRUE Headwaters of the Wolf River all the way downstream as far as Blackjack Road. (Three Bridges Road is the designation granted to that road by the "locals", as there are indeed three bridges crossing the Wolf River on that stretch of thoroughfare).

The day before our personal "First Ascent" of the Headwaters Area - on March 10th, a group which was comprised of three members of the Wolf River Conservancy, and the Bluff City Canoe Club made a similar journey from Baker's Pond, downstream as far as Chapman Road - approximately 4 linear miles upstream of Blackjack Road. (I say "linear" miles, as opposed to actual mileage because ... it seems that the Wolf River and many other meandering rivers actually slither like a snake running from a bulldozer, so - actual mileage traveled in canoeing such bodies of water is FAR GREATER than the distance estimated upon such easy devices as GPS and maps!) These aforementioned guys ranged in age from early-fifties to 76 years old, and their EPIC first ascent from Baker's Pond was an unprecedented journey which included an initial overland drag from the Baker's Pond parking lot of an hour and twenty or thirty minutes to the point at which they also entered Goose Nest Creek and could float in their canoes.

Tim and I instantly came to agreement when he said, "It really doesn't count to ME - if you drag your boat BESIDE the river, as opposed to actually staying in the water and suffering through the BRUTAL Buckbrush". Each person has their own individual belief system, and ways of thinking, so - I will avoid that and any other debates on this matter as they are objectively relative and completely unnecessary... As I said earlier - To say, " I have canoed the WHOLE Wolf River", means many different things to many different people...

On our "First Descent" of the Headwaters Region of the Wolf River - We affectionately named different sections of our EPIC journey monikers like: Ironwood Bottom, Soggy Bottom, The HELL Woods, The Terrible AWFUL, The Snake Farm, and several other names unfit for tender ears/eyes! :-)

We paddled under the Chapman Road bridge at exactly 12:58 PM and decided, with some trepidation, to continue onwards to Blackjack road. This next 4 hours was one of the MOST terrible runs I have yet attempted in a canoe, and - I DON'T recommend that anyone EVER attempt to follow in our paths...

After suffering through nearly 8 hours of Terrible Awful woods, sprinkled with occasional stretches of Beautiful, open water, rocky, mossy bluffs, and more Buckbrush than you can even imagine - we FINALLY came out of the undergrowth directly beside the middle bridge on "Three Bridges Road"... (Blackjack).

I paddled under the bridge, and my friend asked me what I was doing... I HAD to go under the bridge in completion of the Headwaters Area... As I came back under the bridge in an upstream motion - I was never SO Happy to be anywhere, as when we got out of our boats, in completion of the Headwaters Area... You see... This was the ONLY stretch of river I lacked in being able to honestly say, "I have canoed the WHOLE Wolf River". I had, seven years prior, canoed all the way from Blackjack road to the Cobblestones in Memphis Tennessee, so - on this day, I became the FIRST person in recorded history to actually "Canoe the WHOLE Wolf River"! :-)

I am sorry that my commentary was SO long, but I have a LOT to say, and have just begun to scratch the surface... With that said - I bid you adieu!

RIVER ANGELS

On March 26, 2012 at 11:58AM, Jonathan Brown wrote:

Aloha Team!

Hope your Monday has began amazingly!

I wanted to fill you all in on what's been goin on in an effort to communicate updates now so we won't have to during this Thursdays upcoming meeting at Dale's house.

I will be sending an email out before our meeting with our meetings agenda so you can be prepared.

We will be skyping in with Rod as well which will be fun.

Also I think it would be good to broaden our perspective to the main purpose of this event and look at what all OBS has been doing to build a proper event around this most epic paddle.

We have Rod up in Canada plugging in SUPAS every chance he gets and out in the boonies of Canada training for the paddle. We have Dave in the middle of the Pacific on the final stretch of his sail. And last but not least we have the dedicated local paddlers getting out in the middle of swamp and marsh to scout out for the paddle.

I couldn't be more stoked!!!

While the paddle team has been busy getting ready for SUPAS, Rachel and Robinson have been busy putting this event together. Here's and update on what they've been busy with:

THE FOLLOWING IS FOR YOUR EYES ONLY. PLEASE TALK TO RACHEL OR ROBINSON BEFORE YOU COMMUNICATE ANY OF THE BELOW INFO.

- participation:
- Actual participants are still few to none
- Registration system is now up and going. Be sure to direct those that are interested in joining us to this link so they can sign-up, http://standupagainstslavery.com/donate-participate/
- prizes per race/event (first place prizes per event and what will participants be receiving) -
- Outdoors Water Bottles
- Rainbow Sandals SUP Memphis trucker hats
- Still working on "first place prizes"
- additional sponsors and what they will be sponsoring (food, services, financial, etc.)
- HUEYS - $$
- Green Beetle
- Bluff City Coffee - breakfast for paddlers one morning
- Chic-Fil-A lunch or dinner for one event
- Republic - coffee before Father/Son?
- BBQ Shop - food for reception
- Skunx - food for reception
- Maggie Pharm - raffle item
- Grizzlies gear - raffle item
- Southwestern - Wine for final event. high end beer for raffle
- Gingerly Baked - pastries for final event
- local media's involvement (radio, paper, tv)
- Memphis Daily contact through WRC
- final event
- Still needing the permit to be in the park. Starting to think of "Back-Up" plans.
- need to talk to Star and Micey
- Food/Vendors coming together
- Nicci has contact for gates/tables/chairs. Need to connect with her Monday
- Volunteers
- U of M Hockey; UofM Baseball needing to finalize; ATO Fraternity is discussing it at their meeting today; Pike has been contacted; Sports and Leisure Majors will be contacted this week via Jill Creed; A middle school

- not sure which one - will receive some of the flyers this week

JB

PS: thank you Mike for bringing pizza! Thank you dale for your house and hosting this meeting! Let us know if we need to bring anything.

On March 27, 2012 at 10:19AM, John Henry wrote:

Good morning all... My yesterday was joyfully spent upon the Wolf River... Bateman Road to Moscow run... One of MY Favorite places! It was a Beautiful day! :-)

I was brainstorming on the way there, upon the water, and on the way back.

As a group of folks with a common interest - what can we all do to get the word out there, and get not only donations, but especially "People On The Water"? IE : Participation, in terms of participants, for SUPAS will garner us the most exposure/success...

A LOT of logistical planning/implementation has already been accomplished. Most all the minute details have been worked out. We now just need to get boaters signed up, and on the water for at least the 13th, and especially the 14th - the last two days of SUPAS.

Between WRC, the BCCC, and all the other folks we know - it seems like we should be able to get a bunch of boaters together for at least a "Kennedy Park to the Mouth" paddle or race - or BOTH!

The Lagrange mayor agreed to paddle with SUPAS form Lagrange down as far as Leatherwood. My friend Robert Honufsosky is working on having County Mayor Mark Luttrell meeting the team at Mud Island (short of a miracle - AC Wharton will be busy with his Celebrity Golf Tournament on the 14th), and the Germantown mayor meeting the team at St George's on Friday the 13th.

Any suggestions on Press, and TV participation would be GREAT!

Who amongst you knows of ways to get people involved - with a paddle in their hands?

I know that Rachel and Robinson are doing a Great job of pulling this all together, and feel sure that collectively - we can b of further assistance to the SUPAS cause.

I am up for any ideas, and hope that we can make this event a Success that we can ALL be proud of.

On March 31, 2012 at 9:30AM, John Henry wrote:

I understand that after I left the team meeting Thursday - y'all decided to start the SUPAS trip from Baker's Pond after all...

When you reach the confluence of Baker's Pond creek and Goose Nest Creek (after an hour and 15 minute overland drag) and see how exactly much more water runs out of Goose Nest Creek than out of Baker's Pond, perhaps then - you will see why I have repeatedly stated that Baker's Pond is NOT the true headwater of the Wolf River, only the SMALLER of TWO bodies of water which comprise the Headwaters of the Wolf River... Perhaps not.

The elevation of Goose Nest Lake, as compared to Baker's Pond, AND the flow rate of both clearly demonstrates my point, but - it's a group decision, and I will henceforth keep my opinions to myself . :-)

Anyway - I look forward to this EPIC adventure, and am sure it will be one for the books!

John Henry Hyde - SUPAS Advance Scout/Guide in the Headwaters Area :-)

Dale Sanders replied:

Thanks John, we sincerely appreciate your decision to start with us at Bakers pond. I am truly looking forward to paddling (dragging our craft first mile or so) with you and the others. How exciting. You are an amazing guy with so much to offer. Thank you for everything.

John Henry replied:

Actually - Dale, i think that you may have misinterpreted my message. I stand by saying that I will not pull my boat from Baker's Pond based purely upon nostalgia, and misinformation...

I will, instead, meet y'all @ Baker's Pond parking lot on Sat April 7th AM - to load yall's camp 1 gear into my truck (your preference of when), I will then deliver my truck to Camp 1, and head from there w/someone(?) shuttling me to Goose Nest Creek. I will launch as previously agreed (SUPAS scouting trip), from behind Mrs Brown's house on Goose Nest Creek, and meet y'all @ the confluence of Baker's Pond runout and Goose Nest Creek...

I see NO point in an hour drag simply because someone mistakingly called Baker's Pond the Headwaters of the Wolf River years ago! The locals know the truth. I know the truth. It's ALL good - I will simply meet y'all @ the confluence. I feel sure we can work out a mutually agreeable shuttle for launch day. If not - I will supply my own.

Very respectfully yours - John Henry

On April 1, 2012 at 9:10AM, John Henry wrote:

I QUIT!

Hey y'all - I have had a change of heart and have decided to back out of SUPAS completely. I am sure y'all will be successful in all your endeavors, and wish you all the BEST. Sincerely -John Henry

PS - APRIL FOOLS!

:-)

> Y'all ain't gettin rid of me THAT EASY! :-)

Mike Watson replied:

I hope this is an april fools joke!

I wrote the above before I actually saw your PS. :-)

Mike

Dale Sanders replied:

I must admit, that is the best "April Fool" joke pulled on me in long time, must be getting senile. Was suspicious when I saw tie video but didn't scrawl down far enough. Just wait, I'l get you back. My best was getting married to Meriam on April 1st - She is still mad. (Married her in Philippines where they know nothing about April Fool.)

Jonathan Brown replied:

Ha! That was classic!!!!

On April 1, 2012 at 6:45PM, Jonathan Brown wrote:

FIRST OFF:

WELCOME BACK DAVE!!!! YOU WERE MISSED! CAN'T WAIT TO HEAR YOUR STORIES!

This past Thursday, John, Richard, Mike, Dale, Rod, Rachel, Robinson, and Myself got together over drinks and pizza at Dale's house (Thanks Dale and Mike). It was a great time of kicking back, conversation, updates, and planning. Here's the summary:

RIVER ANGELS

VOLUNTEERS:
- *Rachel and Robinson have had great favor in this area. We currently have ATT, Frats, BCCC, and others committed to help volunteer this event.*
- *OBS will also be providing volunteers for our logistics/support team.*

PARTICIPANTS:
- *We may have volunteers but without participants the volunteers would be bored. That's where OBS is working over-time to get people on the water. At the moment the numbers are slim, but we trust OBS is on top of things and will come trough as the doors open.*
- *Please continue to spread the word!!! Direct others to our site and the FB page. Get them to follow the event on twitter (@obs_supas). If you haven't "liked" the SUPAS FB page please do so and feel free to "share" anything stories and updates that come from SUPAS.*
- *Robinson worked with some BCCC members to create a most professional registration system for SUPAS.*
- *If you know ANYONE wanting to jump in the water with us on ANY section (including BCCC members) please direct them to the SUPAS participation page on the website. We need EVERYONE signing up officially so OBS can plan accordingly and have the opportunity to raise funds.*

EVENTS:
- *If you haven't already checked out how the Wolf Sections are set up, please go to the participation page on the SUPAS website and check it out.*
- *OBS is currently working on rounding up high school, college, and Father/Son combinations for our three featured events set to take place starting April 13th-14th. Tentatively, OBS has the U of M Hockey team wanting to join as well as, through John's connections, the Mayor of Lagrange and possibly WC Wharton. Otherwise the SUPAS Team will have to wait and see the fruit of OBS' labor. Hopefully we will have a good turn-out.*
- *The final event at Greenway Park April 14th was at risk due to an over-looked permit requirement, but through John's connection again, the permit was issued! :) Only problem now is the cost. Looks like it will be around $1000 which OBS is actively looking to cover.*

- If all goes well, OBS is planning on having food, live music, local media coverage, games, raffles, and prizes awaiting us as we paddle in on our final day.

SPONSORS:
- We have received much support from local businesses. From printing to water bottles, to our new friends at Aquapac (thanks Dave!), the response has been great.
- Currently OBS is actively working over-time to get sponsors to provide 4, 35 mile Walkie-Talkies for the SUPAS & Logistics Teams as well as grand prize items and finances to cover event costs.
- A company has agreed to do a price match donation of $500. Please let others know and ask Rachel for the name of this company if needed.

MEDIA:
- Rachel is currently trying to connect with Memphis Flyer, Commercial Appeal, as well as local news and radio stations.
- Through John, there is an interest in doing an interview with a local news station.
- Rod just recently was featured in Chatham-Kent News which did a great job writing about SUPAS! Here's the link: www.ckreview.ca/2012/03/chatham-adventurer-paddles-wolf-river
- Luke Short was bale to get SUPAS connected with SUPConnect! Rachel and Luke are following up with this to see what this connection will look like.
- If you have any media connection or opportunities please send this info to Rachel.
- John just got Chris Reyes from Live From Memphis to join us starting Sunday night til new 72.

PADDLE PLANS:
- We will be either staying at Dale's house Friday night or camping in the Baker's pond area. Please clarify this week.
- For those on the team who would like to start at Baker's Pond, we will be starting earlier than our yet-to-be-decided time near Mrs. Brown's house. Please clarify times this week.

*- Rod is flying in this Tuesday and Dale and Rod will visit John Ruskey to pick up the SUP boards.
- Dave will be flying in Friday.
- We decided that waders aren't the best solution, snake guards are recommended but not provided, long sleeve top and bottoms are recommended.
- Dale and Rod will be taking care of Dave's gear and food needs.
- While on the water we will be uploading media and stories to our social networks as well as promotional material for our sponsors.*

If I missed anything please fill us all in!

Jonathan

Meanwhile, on the Bikecar front, I still hadn't found a shipping company for the bike. Dozens of bid requests had been declined, and the few companies that had accepted replied with offers that were well out of our budget range. With only a week to go until the SUPAS team began their descent of the Wolf River, I was genuinely freaking out in my head. My hopes of having the Bikecar in Memphis prior to the April 7th launch hadn't materialized. In seven days, I'd be knee-deep in a snake-infested swamp somewhere in northern Mississippi, far from cell phone range. For my own sanity, I needed to know that before I entered "the most BRUTAL bottom in the south" (as John Henry called it), the Bikecar was at least en route to Memphis. I contacted Paul Adkins, our "bike angel" in Eugene, and assured him that I was working hard to find a shipper. And then, in a last-ditch effort to find a solution, I reached out to my Memphis friends and an organization called Truckers Against Trafficking (a non-profit organization that educates, equips, empowers, and mobilizes members of the trucking and travel plaza industry to combat domestic sex trafficking).

Hi Paul,

I'm in the process of finding a carrier to pick up the Bikecar at your place in Eugene. I need to know if a transport truck is legally allowed to drive on your street, whether there is a weight restriction or not. If so, we may need to meet the driver at a nearby industrial area or even a Walmart store, someplace where a truck can go. Can you get back to me on this?

Cheers,

Rod

Paul Adkins replied:

Rod, No problem for our street. Have them drive up to our house - or have them come up to the last turn off right before our house (Cross St and N Polk St) and that will make it easier so they don't have to do a u-turn on N Polk.

We can push the bike car 3 or 4 houses to get it to Cross St and N Polk.

Paul.

Email to Truckers Against Trafficking:

Greetings,

My name is Rod Wellington. I am a Canadian self-propelled adventurer currently based in Chatham, Ontario, Canada. I am in the process of securing a carrier to ship a four-wheeled bicycle from Eugene, Oregon to Memphis, Tennessee. This two-seater bicycle will be used by British adventurer Dave Cornthwaite in a 1001-mile charity ride from Memphis to Miami. The ride will be raising funds for breast cancer awareness. I will be joining Dave for the first leg of the journey.

Unfortunately, I am having some difficulty finding a carrier willing to ship the bike as is, intact and on wheels, not in a crate. It is my belief that the bike would be best mounted on a flatbed truck. My goal is to find a carrier company or an owner-operator with a flatbed truck that would be willing to transport the bike from Eugene to Memphis. We are looking to keep shipping costs to a minimum. Discounted or donated free carrier service is most welcome. In exchange for discounted or donated free carrier service we can offer prominent placement of the carrier's name, logo and website link on Dave Cornthwaite's website, on my Zero Emissions Expeditions website (see links below) and in online social media prior to and during the Memphis to Miami Bikecar journey. This is a great opportunity to get free publicity for an owner-operator or shipping company.

Any assistance that Truckers Against Trafficking can offer in helping us find an owner-operator or carrier company that would be willing to help us transport the Bikecar would be greatly appreciated.

SUPAS (Stand Up Against Slavery) - Helping to Combat Human Trafficking

On April 2, I will be travelling to Memphis, Tennessee to partake in a historical paddling expedition on the Wolf River in northern Mississippi and western Tennessee. Myself and three others, including British adventurer Dave Cornthwaite, will attempt to descend the 105-mile-long Wolf River from its source in northern Mississippi to its confluence with the Mississippi River near downtown Memphis. Our team will be descending the Wolf River on stand up paddleboards, a feat never before attempted nor achieved. Funds raised from the SUPAS expedition (SUPAS - Stand Up Against Slavery, April 7-14) will help fund the implementation of an aftercare facility for survivors of human trafficking in Memphis, Tennessee. Operation Broken Silence, a Memphis-based collection of abolitionists who are speaking and acting against the growing threats of mass atrocities and modern slavery in our world today, will oversee the aftercare facility. Operation Broken Silence has numerous media ties in the Memphis area and will be fully utilizing these ties in promoting the SUPAS expedition.

It is my hope to open a line of communication between Operation Broken

Silence and Truckers Against Trafficking. Together, these organizations have a unique opportunity to pool together their resources and social networks and move forward with a concerted effort to help raise public awareness of human trafficking and play a role in its eventual demise.

———

Dave Cornthwaite and I will be embarking on our Bikecar journey on approximately April 19. Owing to the fact that we will need to do a tune-up and some minor repairs on the Bikecar, the ideal delivery date in Memphis would be prior to April 15. The Bikecar weighs approximately 220lbs, is 8' 6" (102") long, 5' 7" (67") wide and 4' 7" (55") high.

Feel free to share this email with those that you think may be able to assist us. We welcome all ideas and comments from the North American trucking community. Thank you for taking the time to read this email.

Cheers,
Rod Wellington

SUPAS (Stand Up Against Slavery)
www.standupagainstslavery.com
Operation Broken Silence
www.operationbrokensilence.org
The CARE Network and OBS Safe House
www.operationbrokensilence.org
Dave Cornthwaite
www.davecornthwaite.com
Dave Cornthwaite - Bikecar expedition
www.davecornthwaite.com/#/2012-bikecar-memphis-miami/4561260435
More Bikecar info (Paul Everitt, Bikecar builder - Going Solo blog)
www.goingsoloadventure.blogspot.ca
Zero Emissions Expeditions
www.zeroemissionsexpeditions.com (N.B. Website now defunct.)

Truckers Against Trafficking replied:

Rod,

I think you have a great cause you are fighting for, and I wish you the very best. Unfortunately, we do not work with carriers or owner/operators to transport items. We are a 501c3 charity fighting to end human trafficking by securing the support of the trucking industry to call into the national hotline if they see underage prostitution or evidence of pimp control. I hope you understand. I suggest you contact HR people at trucking carriers and see if they would be willing to help. Best of luck to you in your search for a carrier.

General email blast to Memphis friends regarding the Bikecar:

Greetings folks!

As you all know, once the SUPAS expedition is finished, Dave will be using a four-wheeled bicycle to journey from Memphis to Miami. I will be joining him for the first leg of the journey. The Bikecar is currently in Eugene, Oregon and needs to be shipped to Memphis pretty much asap, certainly by April 15 at the latest. Trouble is, I'm having difficulty securing a carrier service that is willing to ship the Bikecar for a fair price (quotes have ranged from $400 to $4500!). Most shipping companies want the bike crated but I cannot ask the person in Eugene to crate the bike. The frame does not break down and building a crate for the bike would be a job that I don't feel comfortable placing on someone. So far I've contacted about 40 shipping companies and owner-operators. Most are unwilling to ship the bike. I've listed the bike on UShip, which is an auction bidding website where shippers bid on items to be shipped. Lowest bid gets the job. I had one offer of $1500 and one bid of $490. I accepted the former bid, paid a deposit, contacted the shipper numerous times and have heard nothing back. So, time to move on to another plan.

My brother-in-law, who works as a truck driver trainer, suggested that I get a hold of Truckers Against Trafficking, a non-profit organization combating human trafficking at truck stops. They are based in Colorado. Not sure if you folks are familiar with them but they seem to be raising a fair amount of awareness about human trafficking in the trucking industry. Check out the video on the homepage of their website. I think you'll be impressed. I sent them an email asking for their help with the Bikecar shipment and informed them about SUPAS and Operation Broken Silence. Feel free to contact them at www.truckersagainsttrafficking.com. (see copy of the email below)

Another contact that was given to me was Dave Nemo, a Sirius radio personality who hosts a talk-show about trucking. Nemo partners with Truckers Against Trafficking and mentions them often on his show. Nemo is based in Nashville. There may be an opportunity to reach a wide audience regarding the SUPAS event as well as the Bikecar journey and our shipping dilemma. www.davenemo.com

Just received a reply from Truckers Against Trafficking. They're unable to help. Too bad. The search continues...

Any thoughts y'all can lend at this stage would be most welcome.

Cheers,

Rod

Rachel Sumner replied:

Rod,

I'll be thinking about this as the day continues and hope to get you answer ASAP!

I'll let you know.

Rachel

Robinson Littrell replied:

Rod, I believe you're on the right track. Perhaps you could get a donor to help cover the cost? Maybe you could ask the breast cancer charity to help find a donor among their survivors?

Dale Sanders replied:

Rod, I have no clue how top get the BikeCar here. Once here, I can handle all transport around, storage and have plenty of tools for working on it. Just no Idea how to get it here. Wish I could help but shipping is out of my league of knowledge. If I come up with any remote possibility I will immediately let you know.

Jonathan Brown replied:

Hey Rod, I honestly don't know how to help with the shipping. Let me know if you do come up with a solution. In the meantime, I will be thinking of an option and let you know if I come up with something.

Can't wait to kick it via Skype Thursday! Maybe we could make the shipping thing a point to discuss to see if anyone else has ideas.

JB

Meanwhile, Corn's 3156-mile sailboat jaunt across the Atlantic from Mexico to Hawaii had ended and he was eagerly stretching his sea legs on the streets of Honolulu. I reached him via email with some less-than-stellar Bikecar news.

Hi Dave,

Congrats on the completion of expedition #5. Glad to hear it all went well. Looking forward to hearing some great stories from the voyage.

Well, we have an issue with the Bikecar. It's still in Eugene. The carrier who was willing to do it for $500 never returned my emails and calls, even after I put a deposit down on the shipment. Luckily, I was able to secure a credit with UShip for the deposit. I made several other calls to other companies to no avail. Many companies won't touch the bike without it being crated. I have listed it again on UShip. I've received five bids, the lowest is $1270 plus fees ($1288 total). It's your call on how to proceed. Are you willing to pay $1300 to ship the bike? Of course, lower bids might come in. The advert is set to expire April 4 unless I extend it. I've sent off 50 bid requests in the last couple days. I will send off more tonight.

I've been brainstorming with others, but nothing much has come of it, short of finding someone with a trailer in Eugene who just happens to be going to, through, or near Memphis and is willing to help us out for free or a cheap rate.

I fly to Memphis Monday morning. Dale is picking me up, then drive to John Ruskey's in Clarksdale to pick up the SUPs

Dave Cornthwaite replied:

Hey Rod,

How are you buddy? Still doing loads of post sail admin so a little off grid (lovely!) til I fly on Thursday. Sorry for not being in touch sooner.

Dang, not good news on bikecar, what a bugger. I could just about stretch to $500 so $1300 is a no go. Let's see if anything else comes up through uship and if not then we'll have to scratch it.

Really don't want to miss out on this, especially because of the work you've put in. Was looking forward to pedaling out of Memphis with you. Maybe something will come up, if not I'll have to be creative with another journey option!

Thanks for everything on this. Enjoy your week in Memphis, knock em dead with your talks and see you on Friday morning.

DC

Later that day…

Dave,

Just got a message from a carrier offering $400 if I can get the Bikecar from Eugene to Colorado by Friday. A tight squeeze indeed, but at least a positive direction. The trick would then be to find someone to ship it cheaply from Eugene to Colorado. I sent a question asking where in Colorado he plans to be on Friday. Will keep you posted.

Another option: I'd really like to see this journey happen as well. Linking up the Bikecar expedition with the Memphis visit was a brilliant move Mr. C. There have been a good many serendipitous moments throughout the planning of this, beginning with Paul Everitt's trade show offer, and I believe there will be more such moments to come. So, to help make it happen, I'm willing to split the cost of the shipping. Hopefully the bids will be driven under $1000, which will be good for both of us, and we can get this thing rolling. Perhaps in exchange you can do some video editing for me when it's convenient. Thoughts?

Have you thought of storage for the bike in Miami? Somewhere else in Florida? Are you planning to hook up with Justin Riney while in Florida? Perhaps he could be of some help in locating storage options. I'm still interested in using the bike in Ontario at a later date. I'm also open to other options.

Other stuff:

Friday - Will you have time to buy food for the Wolf after you arrive in Memphis? You mentioned a morning arrival. Does Dale have your ETA? There is a good chance that we will be staying at Dale's Friday night and proceeding to Baker's Pond on Saturday morning. (There had been talk of camping at Baker's Pond on Friday night in order to get an early Saturday start, but it appears that the others have decided to drive to the source Saturday morning.) If time will be an issue, Dale and I can rustle up your food for you during the week. Email us a shopping list if you think time will be tight. Also, we are looking into borrowing some lower leg snake protection for you, possibly boots. What size boots do you take?

Cheers,

Rod

Dave Cornthwaite replied:

Hi mate,

I agree, there are surely some more turns to experience in this bikecar adventure before the week is out. Cheers for the offer is splitting costs Rod, let's see what happens this week and make a call on Friday...

If you and Dale have time to get food this week that would be great. I'm not at all fussy, happy to eat what everyone is eating. What's the general plan where that's concerned?

I have no idea how Friday will shape up, so sorting stuff before then will leave us free to discuss the Wolf rather than rushing around markets. Ace.

Boot size, UK - 10. So thats US 11, I think. Yep, let's keep those snakes off!

    Memphis was sunny and warm the day I arrived, a stark contrast to chilly Ontario. Dale Sanders was waiting at the airport's luggage turnstile, slowly pacing with his head lowered, intently checking his phone for SUPAS updates. His white hair was pulled neatly in a ponytail and topped with a baseball cap. As he raised his head, I could see his bushy white beard and dark, searching eyes.

"Heyyyyyy, thare heee izzz!" he squealed, breaking into his signature wide grin. A firm handshake and hug followed. "How waz yer flight?"

We shouldered the heavy portage packs and duffle bags and comically waddled to his waiting mini-van. A three-tiered trailer used for transporting small watercraft was hitched to the rear of the van. The trailer, as I discovered later, had been hand-built by friend and fellow Bluff City Canoe Club member, Anna Hogan.

"Yoo'll meet Anna at the Canoe Club meetin' day afta t'morra," said Dale in his southern tongue. "Evvabuddies lookin' fowad t' yer presentation."

I smiled wide. It was good to be back in Memphis.

We drove 90 minutes south to the town of Clarksdale, Mississippi. Clarksdale is not only home to Robert Johnson's famous "Crossroads" location, but also the Quapaw Canoe Company, owned and operated by river legend, John Ruskey.

"Welcome to tha Quapaw!" shouted John as we emerged from the van.

"A pleasure to finally meet ya, Rod!" said John as we shook hands and shared smiles. "And good to see *this* youngster again! How are ya, Dale?"

The three of us spent a few minutes swapping paddling stories and discussing SUPAS details before John gestured toward the Quapaw building, "C'mon. Let me show ya around!"

The sprawling, three-level Quapaw Canoe Company building sits on a huge lot in downtown Clarksdale, a stone's throw from the Sunflower River, a tributary of the Mississippi. Nearly every building in Clarksdale has seen better days. A blanket of unwashed poverty shrouds the town. Vacant storefronts yawn and show their empty innards. Their withered exteriors are long faded and peeling. Dark grey woodgrain peaks through the cracked paint. A stroll down any street is a stroll back in time. Sometimes, the place looks like it's stuck in 1962. It hasn't aged well, that's for sure. Here, in the birthplace of the blues, one gets the impression that life will never get better because it was never all that great to begin with. The poverty, the desperation, the blatant racism, the continued segregation, the shameful past, the wilted present. With so much stacked against it, Clarksdale seems destined to fail and fall to ruin. But like similar towns all across the South, Clarksdale prevails. It refuses to fold. As long as there is life, there is blues. And as long as there is blues, there's an audience.

Each new month brings a new music festival. Tourism dollars bolster the town's economy. What Clarksdale lacks in aesthetic appeal—although, its weathered appearance *is* an honest aesthetic—it more than makes up for in cultural treasures. Whether it's listening to a grizzled bluesman strum out a sad lament on a hand-crafted cigar box guitar, or sweating out a fiery tamale on the dancefloor at Red's, or squeezing into a packed room to catch a rising star at the Morgan Freeman-owned Ground Zero nightclub, or standing in the footsteps of Robert Johnson at the famous Crossroads, Clarksdale, socially and artistically, has plenty to offer.

The cracked, patched, and paved sidewalk out front of the Quapaw Canoe Company building gives way to a large, empty parking lot. An ancient, flatbed work truck, its original firetruck red paintjob now a dull pink, sits idle near the roadside. The yellow brick building, topped with an auburn-coloured metal awning, looks positively industrial—more in line with a used car lot than a canoe builder's business. But a 15-foot-high teepee—sans the canvas wrap—and a smattering of colourful hand-painted murals shows there is something wonderfully creative dwelling beneath the generic façade.

The building is split into four distinct areas. The retail section of the building features a glassed-in storefront. (On this day, the 14-foot SUP that Dave Cornthwaite used on his Mississippi River descent was on display.) Behind the storefront is a large storage area that houses a sizeable cache of paddling-related gear and a fleet of rental canoes, kayaks, and SUPs. These boats and boards are also used on guided daytrips and multi-day expeditions. A fully equipped workshop, where large cypress strip canoes are skilfully assembled by hand, occupies one quarter of the upper level while a two bedroom apartment dubbed "The Owl's Roost" occupies another quarter. John's office and library, colourfully decorated with an impressive assortment of river treasures (driftwood, rocks, and fossils), is located on the lower level. Overlooking the Sunflower River is an open air, warm weather workshop surrounded with rusting antiques and other river detritus.

A covered parking area houses two large cypress strip canoes, both built here at the Quapaw. Affectionately named Grasshopper and Junebug, the 25-foot boats are used for group tours and school programs. Educating people about the Mississippi River is the Quapaw's main goal. The subject is at the heart of John Ruskey's passion for people and rivers. He loves bringing them together. The Mississippi, in John's eyes, is a national treasure, and should be revered as such. He shows people firsthand the advantages of travelling by river, especially when passing through stretches of industry-free riverbanks that John has dubbed the "Wild Miles." His business provides employment opportunities for local youth, much needed in a state known as the poorest in the nation. As a musician and artist, he contributes greatly to the arts community in Clarksdale and lends his hand to music festival planning and promotion. In so many ways, John Ruskey truly *is* a River Angel, both on the water, and off.

Dale and I returned to Memphis with four SUPs strapped to Dale's boat trailer. During the drive north, Dale quizzed me about past expeditions: what gear I used, why I chose certain routes, the challenge of being an author, and the ups and downs of travelling the length of the Mississippi River with my old friend, Scott McFarlane. In turn, he told me about his Navy days and the many countries he'd visited with his wife, Meriam.

"She'll be home when we git thare," said Dale as he drove. "She's lookin' forward to meetin' ya!"

The long driveway to the Sanders' home cuts through a small forest of lofty maples and passes by a pond, a chicken coop, and a fenced-in garden.

"Darn deer 'round here were eatin' up our lettuce," said Dale, pointing at the garden to our left. "Had to put up that wire fence to keep 'em out. I'd shoot 'em all dead, but my neighbour thare, he's a doctor, and he don't like the sound of guns. People 'round here think

I'm a little crazy as it is. Don't need to give 'em any more reason to think worse o' me."

Dale may be a little loopy at times (it happens when you get to be his age), but I certainly wouldn't label him an eccentric. His spelling is atrocious, his memory isn't the best ("What day of the week is it?"), and he may seem weird to some of his neighbours, but I've seen a lot worse. I haven't spent much time hanging out with him, so I can't say I know him well, but one thing I *can* tell you: he has good taste in houses. He may call himself "The River Rat," but the truth is, he lives in a *mansion*.

The city of Bartlett lies on the outermost ring of suburban Memphis and is eloquently dotted with stately country homes on generously sized lots, many of them graciously wooded and preciously manicured. The Sanders' huge home, three levels of spacious loveliness wrapped in a butter and sugar white-coloured wooden exterior, was just as pretty on the inside as it was outside.

We were greeted at the garage door with barks and howls from Molly, the family's aging, but sprightly, miniature schnauzer. Dale returned the greeting with howls of his own. I got the feeling these two did this every day.

As we climbed the stairs to the main level, Dale pointed out a signature on the landing wall.

"Betcha recognize that name," he said.

The sole signature on the wall belonged to Dave Cornthwaite. It was dated September 2011.

"Dave stayed with us for a couple days after his SUP trip down the Mississippi," said Dale. "After we finish the Wolf, I'm gonna have every member of SUPAS sign this wall. Maybe it'll become somethin' like a paddler's hall of fame, or maybe a paddler's *wall* of fame. That is, if Meriam doesn't paint over it first. She's thinkin' blue would look nice on these walls. I'm gonna hafta talk her outta that notion."

We emerged from the staircase into a large living room flooded with natural light. A warm spring breeze brought the soothing tone of oversized wind chimes in through a half-opened balcony door. A collection of blue glass bottles, cleverly arranged in descending order by height, occupied a ledge above the door. Soft yellows and whites coated the walls and vaulted ceiling. Above, a spacious hallway led to the upstairs bedrooms.

"Meriam! We're home!" shouted Dale.

"I know, Dale," she said, nonchalantly. "I heard you barking downstairs."

Dale turned to me and smiled.

"Come meet Rod Wellington," he said. "He's gonna be stayin' with us for a few days."

Meriam greeted me with a wide smile and a warm hug. She was a slender, attractive woman in her early 50s, at least two decades younger than Dale. Her thick, straight black hair hung neatly past her shoulders and a leftover accent from her Filipino heritage still lingered when she spoke. She and Dale met in the Philippines while he was stationed there with the U.S. Navy in the late 1970s. They married, travelled extensively, settled in Memphis, and raised three children. She is actively involved with her church group and regularly serves as a volunteer with events and gatherings.

After our short visit, Dale led me back downstairs and showed me a room that can only be described as a nautical-themed man cave, complete with a huge, manly, wall-mounted TV and a huge, manly fireplace. The walls were decorated with hefty rope and fishing net floats. An antique diving helmet topped a small end table and a large wooden ship's wheel hung above a sofa bed that faced the TV. It was the kind of room where a manly man could exist for decades, as long as someone delivered him food and water.

An adjacent room was less cluttered. A box of toys revealed that it was probably a rumpus room for the Sanders' grandchildren. Another sofa bed backed against the far wall and a few armchairs completed the décor.

"These rooms are for yoo and Dave. Yoo guys can choose who sleeps whare. The pull-out couch is plenty comfy. Meriam's left ya'll summ beddin' thare. Tha fridge is in tha garage, jus' outside that door. TV ova thare. Hot shower in tha bathroom. Y'all can use our towels, or use yer own. Don't matta. Make yerself at home! Anythin' yoo need, yoo jus' let me know, alright?"

I thanked Dale for welcoming me into his home and for providing a place for my father, sister, and brother-in-law when they arrived in a week's time. Although the term had yet to find its way into the Memphis paddling community's vocabulary, Dale, Meriam, and many others that I'd meet in the coming days, all qualified as genuine "River Angels." It was an honour to be in the company of these generous souls.

Of course, every shiny coin comes with a flipside, just as every pretty thing casts a shadow. If I was going to savour the sweet, then I was going to have to suffer through the sour as well.

Meriam's rigid adherence to her faith left me feeling uncomfortable. Crucifixes and other church-related paraphernalia—a framed Bible verse here, a Christian self-help book there—were evident throughout the house. I silently hoped our conversation would not lapse into vapid sermonizing or some twisted moralistic exhortation. My soul, as it was, needed no rescuing. Thankfully, sacred ground was rarely treaded upon during our short talk. When it was, I tactfully kept my opinions and judgements to myself.

In all the travelling I've done in Canada, in all the homes I've stayed in, lived in, and shared with other people, the topic of religion rarely came up. I can only remember one time dealing with the same uneasiness I felt in the Sanders' home that day. It had something to do with a housemate's girlfriend who regularly took it upon herself to mingle with prostitutes in downtown Vancouver, B.C. and persuade them, through the word of God, to get off the streets and make something of their lives. That scenario, and my unfortunate proximity to it, did nothing to dissuade a long-rooted distrust of anyone who chooses to have faith in imaginary deities. Frankly, people like that scare me. I don't mind distancing myself from a person who wears horse blinders. The only time they notice me is when I'm standing in front of them, when I've entered their narrow periphery. Here's a dose of reality for those people: You're not a child of God. You're just a hairless monkey in a cotton suit. Get used to it.

During my 20 years in Vancouver, I shared time with punks and anarchists, two groups predominantly opposed to the Catholic Church.

I gravitated towards people I shared similar interests with. Because of that, I rarely came in contact with church-goers. Followers of religion had no place in my life, and I had no place in theirs. Until now.

By descending into the American Bible Belt and embracing a community of Tennessean paddlers (and, subsequently, Tennessean *church-goers*), I had stepped well out of my religion-bashing comfort zone. It seemed that *I* was now the minority. That simple fact had me feeling very uneasy.

I'm a polite person. I don't bring race, religion, gender, or politics into general conversation. I try my best not to be judgmental. Many people within the punk scene were able to articulate what I had felt in my heart for many years. When I heard their lyrics in music and read their words in books, their ideas resonated with me. It was a confirmation of what I already knew. At that point, I realized I wasn't alone; I wasn't fucked up. I had found my tribe, and with that discovery came a certain amount of contentment. Yes, I was different from the status quo. My dress and hairstyle went against the grain. I had been a rebel since high school and made no effort to change in order to fit in. Interestingly, I never really felt that I was fighting against "the system." I didn't openly protest against church or government. I simply didn't give a fuck. And before you write me off as apathetic, please know that my "not giving a fuck" was actually not so much a stance as it was just a reflection of the real me. I care about a lot of things, but I don't care about church or state. Those things could disappear and I wouldn't lose any sleep. They are unnecessary in our modern world. Just as the military is unnecessary. As is the public school system. And corporations. And Big Pharma. And all forms of "authority." Fuck the police. Fuck the judicial system. Fuck all human-created laws. It's all unnecessary. The only "laws" that apply to me are natural laws. Why? Because humans did not dream up natural laws—other than recognizing that they exist, and then labeling them as "laws" so the concept could be shared and communicated among other humans. In many respects, I fucking hate humans. I'm embarrassed to be one. But I can't really be anything other than a human (without dying, that is), so, regrettably, I have to live with it.

I once listened to a radio program about a woman who chose to live her life as a cow. She explained that she felt closer to cows than she did to humans. I thought that was a pretty punk rock thing to say. She broke from tradition and became what she was: a human-rejecting cow-lover. There's nothing wrong with that in *my* eyes. She wasn't a *rebel*. She was simply being *herself*. She seemed comfortable in her own skin and was proud of the decisions she'd made and the direction she'd chosen. Really, who can fault her for that? Isn't that what we're all pursuing anyway—*authenticity*? It seems to me that we all want to be comfortable being ourselves, but few of us are. Why? Because we're trying too damn hard to be humans. Fuck that shit! Trying to be human is a waste of fucking time. Just be yourself—free from fucked up human concepts of human righteousness. Nature doesn't give a fuck about you. Nature can strike you down in an instant. But will it? Not likely. Why? Because *it doesn't give a fuck about you*. It never did and it never will. Get used to it. Nature doesn't give a fuck about your religious or political beliefs. It doesn't care about your "concept" of nature. It doesn't care about the Kardashians, or cancer, or how many black kids got shot today by white cops in some fucked up neighbourhood in Los Angeles. IT DOESN'T GIVE A FUCK. It never did and it never will. Get used to it.

On April 3, 2012 at 6:48AM, John Ruskey wrote:

OARSOME DAVE!

Thanks for the beautiful image of being in the middle of the ocean.

Dave, I am back at home base. Yesterday evening I sent 4 yaks back to Memphis with Dale and Rod. Good team-meates you got there!

I am assuming that you didn't want your yellow 14' board since you didn't specifically request it.

As has been discussed, the yaks will be perfect for the rigors of a shallow river. Also, sent our basic rental paddles. Nothing fancy, but you can beat the shit out of them. Let me know if you need anything else.

John Ruskey
Quapaw Canoe Company

On April 3, 2012 at 7:49AM, John Ruskey wrote:

Good Morning SUPAS-ers!

Hey, I just wanted to reiterate to all something I shared with Dale and Rod when they picked up the Yaks yesterday:

Please feel free to adorn these Yaks with any signage, signatures, notes, paintings, decals, doodles, etc, etc that you are inspired by. Any sponsor logos. Anything you feel like pasting or writing on. Permanent magic marker. Whatever.

I know they are going to get beat up on this kind of journey, use with impunity and don;t worry about any abuse. They are tough, that's what they're built for. Its all part of their growing up — "character enhancement."

These boards are beginning a new journey, and it is an honor to have them paddled by such fine gentlemen as yourselves.

Many blessings for your journey.

John Ruskey
Quapaw Canoe Company

On April 3, 2012 at 7:54AM, John Ruskey wrote:

Hey Rachel, Jonathon:

I sent 4 boards + 4 paddles with Dale and Rod yesterday. That was the count I had my crew prepare for before I left for my 2-week Circumnavigation of St. Louis. All others are out on the water, and we have commitments for festivals we're attending next week (Juke Joint - Clarksdale/ Naturefest - Mississippi Museum of Natural Science)

I know that you may be looking for extras now.

I talked to YOLO boards yesterday and we're going to see about getting more boards up to you, if you still think you need more.

Please keep me up to date about what you decide about this. I know how expeditions go, and plans change along the way.

Just keep me tuned in and I will do my best to accommodate!

John Ruskey
Quapaw Canoe Company

Dale Sanders replied:

Thank you from the bottom of my heart John, Believe me the pleasure was all ours. You, your place and Clarksdale were all first class. A paddlers dream. Rod make several references to how he felt in your environment as compared to rememberable life's experiences of his past. You and your programs are special in our hearts special. No doubt the boards will have their own little special markings when returned, including, I hope, signed notations from the paddler. As mentioned, if you Accountant needs further documentation, please have him call me. Thank you so much.

It was pleasing to see that the SUPAS plans were progressing well. Over in Eugene, Oregon, however, things were almost at a standstill.

The Bikecar was still idly occupying a corner of Paul Adkin's backyard and our UShip advert was set to expire the following day. The ad had collected a handful of bids—most of them preposterously inflated—but the lowest, at $1270, seemed like our only sound option. The chance of the Bikecar arriving before we departed for the Wolf seemed slim, but I held fast to the hope. Acting quickly would increase the odds. I contacted Corn in Hawaii and shared my thoughts.

Dave,

I say we try to nail down the Bikecar shipment tomorrow (Wednesday). Lowest bid so far is $1270. We will need to act very soon to secure that bid. I know it's a lot of money, but the carrier may move on to other jobs and may not be in the Eugene area in the next few days. I'll have time Wed morn to deal with this. Your thoughts?

Cheers,

Rod

Corn replied:

Hey Rod,

How's the Memphis crew?

Hmmm. $1270 is a lot. Even half of that is pushing it for me, and although I'm super stoked you offered to help I'm concerned that you're not going to get value for your dime mate. I do appreciate, but I worry!

Let me put out a teasing Facebook post now, suggesting that perhaps the Bikecar journey will be cancelled unless we can find someone to ship it ASAP. That will generate some interest, hopefully.

There's also the danger that once it arrives in Memphis it might not be in the best condition - so there's an element of risk to paying so much money with no time to make it road ready after the paddle. What do you think about this?

I agree though that if we're going to go for it, now is the time else the chance of any carrier disappears.

Let's see if Facebook gives us any leads...

DC

My reply:

Another option to recoup money would be to sell the bike at a later date.

Corn replied:

Some other shipping options popping up on Facebook. Let's have a skype in the morning, Hawaii time - if you're around.

Cheers buddy

Dave

I pitched another idea, something I wished I'd thought of weeks prior.

If you havn't yet, one of us should write up a posting on Craigslist asking for assistance on getting the Bikecar from Eugene to Memphis. My suggestion is to have a post on the Eugene page and the Memphis page.

We still have the backup option of using a carrier, but it's an expensive option. We may have to bite the bullet on this one.

We can still do a Skype chat this afternoon. It's now 1:30pm Memphis time. I will most likely be at Dale's until 4pm.

Cheers,

Rod

Corn replied:

Hey Rod,

Been rushing around like a mad dog. Should settle in the next hour for a chat. Haven't done anything Bikecar related yet, if you have time to shoot out a craiglists post then cool. Will be on Skype very soon.

Assuming there has been no change on uship?

Corn then sent a general email to the SUPAS team. Some good news for a change—his flight details were finalized.

Hi guys,

I arrive into Memphis at 7.13am on Friday morning, Delta 1820 from LAX. Unless you guys are cruising nearby for a pick-up let me know the plan...

Looking forward to seeing you all!

Dave

Our final exchange laid bare the final realization: our time of waiting had expired. It was time to bite the shipping bullet and pay up. Onward…

DC,

Crazy busy at this end too. Dale and I are leaving here in 20 minutes to go to Bluff City Canoe Club meeting and presentation tonight so I'm somewhat hooped for doing the Craigslist stuff until later tonight. Stopped by Bike Plus, local bike shop. They will do required bike maintenance. They said to try carriers ATC and Oregon Express. Sometimes state carriers will ship to a central location, then another carrier will take the shipment onwards from there. Oregon Express may be able to do all the logistics for us. There was a new bid from UShip - $1600. Lowest bid is still $1270.

Chat soon,

Rod

Corn replied:

Ok mate,

I'm off grid now until I land in Memphis. Still wrapping up post sail stuff and trying to see a bit of Hawaii while I'm here. Such a bugger the lower priced courier didn't come through...I can stretch to $700 (a real stretch) but am concerned about having anyone else pay for the rest, it doesn't seem fair!

Could possibly sell off at the end but that's also extra hassle. Maybe I charge a taxi fare every time picking someone up!

No time to do anything now. Will have to risk losing the 1270...see you Friday

Dave

    I gave a slideshow presentation that night for the Bluff City Canoe Club. It was a bit of a fiasco. I talked too long, tried to cover too much ground, and showed way too many photos. The crowd seemed to enjoy it. They clapped and laughed loudly many times—proof that I was doing something right—but I lost track of time and dragged the event past its allotted schedule. Next thing I knew, I was in the parking lot saying goodbye to everyone.
    Fame, the vain pursuit. Expectation, the unripe fruit. Disappointment, the inconvenient absolute. Reality, the threadbare parachute. Reflection, the overdue reboot. Quiet comes a time, when truth it does unwind, a spring too tightly wound, a clasp too tightly

bound. I breathe. I stretch. I walk. I eat. I talk. I smile. I openly greet those who openly welcome me in, so the imagined reinvention can be reimagined again. I clutch. I grip. I grab. I grapple. I seize. I snatch. I take. I tackle the tests that I've failed far too often, and feel my resolve begin to soften like muscles unused and thoughts well abused and excuses excused as I try to diffuse a guilt self-imposed, a reiterated prose that I'm sick of rehearsing and reversing and conversing with no one, none other than me, the chief absentee. To my ego I plea, "Please hand back the key to the place I once dwelt, to the love I once felt, to the heart that won't melt." At the pew bench I knelt in prayer and repose, stilled, calmed, and quieted, a need then arose. I arrived in the present and enacted a plan. Running toward something, toward anything I ran. I dig in my heels. Inventory the feels. Evaluate the reals. And inflate my soft wheels. I roll soundly forward, to the horizon I race. Tomorrow's elusive, it's today that I chase, down alleys and avenues and streets yet unnamed, pursuing a vanity that I once deemed as famed. But now I see different, I'm wiser and wrinkled. Truth, like sugar, is something best sprinkled on days of the week when the outlook is bleak, when efforts seem meek, when hate sometimes speaks in words laced with meanings we solemnly seek. Truth, it seems, is never far from our grasp. But truth, like freedom, is something we abandon on the doorstep of the present. We go off in search of a doubtful reality. Sometimes, we find exactly what we are looking for.

April 4, 2012

"How goes the shippin' battle?" asked Dale.

"Everything seems in place," I answered. "Now we just sit back and wait."

"Good ta heer," said Dale. "Say, how 'bout yoo and me take a drive out ta Baker's Pond an' have a look around. As yoo know, John Henry's claimin' that the real source of the Woof is a small lake on private land. You might wanna see it for yerself, firsthand, an' come to yer own conclusion. Yoo up fo' takin' a drive out thare?"

"Sounds good to me," I said.

"Wail, thare ain't no time like tha present, is thare?" said Dale, smiling.

The 90-minute drive to Baker's Pond took us over the Highway 72 bridge, the all-important destination on Day 3 of the SUPAS expedition. An open area north of the bridge served as a launch site for self-powered watercraft. The SUPAS team planned to overnight there.

As Dale's van sped across the bridge, I got a quick look at the Wolf. It was narrow, muddy, and showed little signs of movement. A short, straight stretch of river came out of a stand of towering cypress trees to the right, and disappeared into another stand to the left.

In the vicinity of Baker's Pond, Highway 72 runs atop a low ridge that forms the extreme edge of the Wolf River watershed. About two miles past the Baker's Pond turnoff, we arrived at a large property parallel to the highway. A house was set back from the road and a small pond occupied a corner of the huge yard. Beyond the house, out of view, was a larger, man-made lake. A water control structure had been built in an earthen dam at the southern end of the lake. The structure controlled the lake's outflow into a dredged creek. This creek flowed south for approximately 800 yards, eventually merging with the outflow from Baker's Pond. The creek, and the private lake, remained nameless until John Henry came along. John discovered that the lake and creek were at a higher elevation than Baker's Pond. He then deduced that the lake, and *not* Baker's Pond, must indeed be the true source of the Wolf River. He christened the lake and waterway Goose Nest Lake and Goose Nest Creek. There is a public road that runs parallel to a portion of Goose Nest Creek and it was here that John planned to start paddling on Day 1 of the SUPAS expedition, in defiance, of course, to Dale's plan to begin from Baker's Pond. The Wolf River Conservancy—of which Dale was a volunteer—claimed that Baker's Pond was the river's source. Dale, perhaps thinking it best not to taint the expedition by starting from somewhere other than the alleged source, chose to stick with his plan of launching from Baker's Pond.

It all seemed so petty: grown men bickering about the source of a river. Was it really that important? Apparently so, because both John and Dale were very adamant about it. Some might say *obsessed*. But, before I heap too much criticism on them, I should note that there are far more detrimental things that they could've been focusing their energy on. They both loved the Wolf River. It was part of them, part of their *being*. They knew it intimately, like a lover. To them, the Wolf was a lover worth fighting for.

Baker's Pond lies in a sparsely populated corner of Holly Springs National Forest in northern Mississippi. The tree-ringed pond is about 300 feet in diameter and is accessed via a one-mile-long walking trail. The pond's surface was calm the day Dale and I visited. The surrounding forest of birch, elm, and oak was beginning to leaf up. The SUPAS team planned to pass through the upper Wolf before the spring foliage fully emerged, hence the decision to begin in early April. Doing so would make navigation easier and would assist in finding channels of flowing water if we needed to portage overland.

Dale and I examined the proposed launch area at the north end of Baker's Pond. The ground there was solid and leaf covered, but excessively muddy near the pond's edge. We walked to the pond's southern end and examined the outflow. The stream was no more than three feet wide and about 12 inches deep—hardly enough to float a SUP. I peered downstream and watched the clear water disappear around a tight bend. The dense forest seemed to swallow the stream. I knew from satellite images that the first 500 yards would be a laborious portage through knee-deep mud and waist-high swamp water. And then there were the snakes. They were damn near invisible. If the weather warmed up, they'd be out in droves. Somehow, our team would have to steer clear of them, nearly impossible when we were passing right through their dens. I cringed, grimaced, and turned to look at Dale.

"Beautiful, innit?" he said.

I was speechless. Nothing here seemed overly appealing to me.

We departed Baker's Pond and drove a few miles west to Chapman Road. A mile on, we came to a low bridge crossing. Dale parked the car and we stood looking at an impenetrable forest that nearly touched the bridge. The river here seemed murky, languid, and wholly uninspiring. It was obvious that in this section of the upper Wolf, the meaning of *river* and *swamp* were interchangeable.

Dale broke the silence.

"That's ware we came out," he said, pointing in the direction of the forest.

My face screwed up in a puzzled look.

"Came out of *what…?*" I asked, scared of his answer.

"Outta tha swamp," said Dale. "That's all swamp in there. Thick, god-forsaken swamp for miles in all durreckshuns."

"Christ!" I thought to myself. "What have I gotten myself into this time?"

My expression must've been evident, because Dale turned to me with a wry smile and asked,

"Havin' any regrets?"

"Maybe…" I uttered.

I watched his smile melt away as the realization of the hell he was sure to encounter downstream of this bridge sunk in. Dale had already paddled and portaged from Baker's Pond to Chapman Road Bridge. He, Mike Watson, and Richard Sojourner had struggled for hours just to get to this point, about four miles downstream of Baker's Pond. *He* knew what the river and swamp were like upstream. *I* didn't.

From my computer desk in Ontario, everything about this expedition seemed straightforward and fairly easy. We enter the swamp with maps and a GPS and emerge unscathed downstream. Really, how hard could that be?

Well, I was no longer viewing a satellite image of the Wolf River on a computer screen in Ontario. I was standing above the river on a bridge in some remote area of northern Mississippi. Beside me was one of only five people who had ever paddled upstream of Chapman Road Bridge. Dale was a tough old coot, that's for sure. But on this day, as we both stared straight ahead into the dense forest, a twinge of fear crept over our bodies. It was a fear we could not turn away from.

"This ain't gonna be no picnic, Bubba," said Dale as he slowly shook his head. "This ain't gonna be no picnic."

SUPAS – Day 1 – April 7, 2012

"Hay, Rod!" shouted John Henry as I paddled around a tight bend and glided up to the flat riverbank where he stood. "Ya gotta keep up, brother, or yer gonna git left behind!"

The others were there, sitting on their boats and boards. They looked impatient, grumpy. They were eager to move on. This wasn't the first time they'd waited for me today, and it wasn't the first time John had dealt me a stern admonishment. After the second such warning, his tone had become more serious, more direct. He'd voiced this one in front of the others, and his words produced a flame of anger inside me. I felt embarrassed and insecure. It was only the third hour of Day 1 and already the resentment and personal grudges were taking hold. With each look of impatience, with each public scolding, with each unspoken complaint, I silently drove a wedge between myself and the others. The comforting idea of *us* had been abandoned at the outflow of Baker's Pond. That was the moment things got difficult. That was the moment when pond became river, and river became swamp. That was the moment when the horizon diminished and tolerance dried up. That was the moment when I had to step off the paddleboard for the first time and drag it through dead grass and sloppy mud. That was the moment things got *real*. The picnic, if ever there was one, had ended. From here on out, escapism became the motivator.

John Henry splashed his way through a pool of shallow water and stood at the bow of my paddleboard. He took a quick swig from his bottle of Guinness stout and spoke directly at me.

"Look, I know it's hard goin'. And I know yer not enjoyin' this—none of us are—but I gotta ask you to pick up the pace a little. I don't havta tell ya that it wouldn't be no fun spendin' the night out here by yerself."

"It ain't no fun *now*," I said, sourly.

"You got that right," said John with a knowing nod. "But lissen, we gotta keep movin' if we're gonna reach James Lowry's by nightfall. Ain't none of us wanna sleep in our boats, so let's keep the pace goin' good, all-right? This swamp's a big, ol' mutherfuckin' mess out here, but we got good maps, balls o' steel, *and* beer! And as long as we got beer, we can git thru *anythin*!"

It was hard not to smile when John Wayne—I mean, John *Henry*—opened his Southern maw and unleashed some salty enthusiasm. When he wanted to, he could downplay even the most serious of subjects. He had a knack for getting under your skin *and* making you feel like a million bucks, often at the same time. He was equally abhorred and adored. He was a hero and a villain, a sinner and a saint, an am and an ain't. He was the crack in the paint, the shine in the varnish, and the unwelcomed tarnish on a tea set. He was a unique fucker, that's for sure. Somehow, he'd found his way into my life, and me into his. Like it or not we were stuck with each other, at least for today. But today would end. Tomorrow, he'd be shuttling our gear with his truck and spouting fire elsewhere. That was a welcoming thought, and I looked forward to the reprieve. Tonight I would go to bed plenty ticked off at the dressing-down he had spit my way earlier. I had no plans to forgive him for that. He and Dale had rushed me when I felt it unnecessary. I had volunteered to properly document the expedition with photographs and video and I needed ample time to do so. Unfortunately, I wasn't given time to do either. From the outset, the expedition was flawed. I'd involved myself with two maniacal manipulators who hated each other and seemed bent on creating a needling hell for one another. Nothing I did would satisfy them. I had trapped myself between two polarized combatants in some fucked up obstacle course disguised as a river. I'd been poked, pushed, prodded, and provoked. I'd been hurried, hastened, and hustled along—*all* against my will. I was *one unhappy motherfucker*.

The quickest way to get me to isolate myself is to rush me. If you do that to me, I will hate you forever. And each time you rush me, the hate compounds. You will stay on my shitlist for life. My shoulders may sag under the weight of my grudge, but my legs are strong and my resolve stronger. I can carry that shit for years without breaking a sweat. You push, I walk. You push harder, I walk further away. I cannot be coached. I cannot be urged. I cannot be persuaded. I am entrenched in my own belief and I believe you will always remain buried beneath me. I don't do deadlines. I do *life lines*. Get used to it or get the fuck away from me. There will be no compromise. *Ever.*

With two days to go until the SUPAS launch, JB sent out his final update. I felt a bit sad that I wouldn't be receiving any more updates from JB. They had been well crafted and informative. It was a sign that the planning stage was winding down. Soon, the action stage would commence. We were all multi-tasking, busy with our side projects. Soon, we'd set all that aside and board our crafts for a great adventure. But first, there were details to attend to—in fact, far more than I bargained for.

On April 5, 2012 at 2:04PM, Jonathan Brown wrote:

First off, keep in mind this is going out to others that may not need to be included in our replies.

Aloha all you beautiful Memphians, Adventurers, and Fighters of Injustice!!!!!! This will be the last SUPAS update between now and the start of SUPAS.

The purpose of this update is to clarify our plans and logistic details.

I don't have an update concerning the event itself. At this moment there are no participants signed up but Rachel and Robinson are working hard to bring everything together. Worst comes to worse we can turn our volunteers into participants!!! :) Also, Rachel is working hard to get the media spotlight on this event which will, in turn, create a ripple effect that will last.

As we paddle we will be documenting the trip in various ways and for various purposes. In the evenings some of the captured info will be uploaded to the SUPAS social network and to SUPConnect. Rachel, Robinson, and SUPConnect will ensure that our stories, photos, and videos are posted, tweeted, and shared properly.

DALE & JOHN= Can you let OBS know ASAP what days and where we will need the help of a logistic team? My thinking was having them for the first 3 nights then again one evening between Tuesday and Thursday, then one more time Friday night. What's your thoughts?

Below is the rough version of our itinerary. Please fill in the blanks and make corrections where needed.

Friday April 6th (Dale's House)
- Arrive at Dale's house no later than 9pm (let Dale know if you're coming early)
- All Paddle Team meet at Dale's House. (We need everyone)
- Final pre-paddle communication
- Equip boards and canoes with sponsor stickers and decorate the SUPs (per John Ruskey's permission)
- Gear and Equipment checklist
- Pack boards, canoes, and gear.
- Sleepover at Dale's!!!
- SUP Colors= Dave/Yellow; Rod/Green; JB/Red; OBS & Luke/Blue

NOTES:
RAY & MIKE= WIll you be dropping in?
LUKE= We will need you at Dale's so we can all leave together Sat. morning. If you have hats or stickers, bring em! Also, I will be bringing your Aquapac bag Friday night.
OBS= It would be great if both Rachel and Robinson show up.
OBS= We will need any sponsors promo materials at this point (stickers, signs, etc.)
OBS= Will you have any OBS promo stuff?

Saturday April 7th (First Day) (Mile Marker 0 to James Lowery's House)
- 5:30am Leave Dale's House

-7am Arrive at Baker's Pond
- Richard and John will help the rest of the team get started from Baker's Pond
- John and Richard will take our first nights gear to James Lowery's House (First Camp Site)
- 9am (?) Meet John just below Mrs. Brown's House
- 6pm (?) Arrive at James Lowery's House
- Luke Short goes back to school.

NOTES:
LUKE= Will you be arranging a ride to pick you up from James Lowery's house?
OBS= Gift for James Lowery?
JOHN= Will you be taking Luke's SUP to David Houston's for Robinson?

Sunday April 8th (Easter Sunday) (James Lowery's House to Pipeline 5)
- Get up for your own personal Sunrise Service
- Leave James Lowery's House at (insert time here)
- John will take our gear to David Houston's House
- 6pm (?) Arrive at Pipeline 5 (Second Camp Site)
- David Houston, John, and Robinson will bring our gear, John's Canoe, & the SUP for the night.
- We still need someone to camp with us who will also take our gear to Dave Houston's House Monday morning.

NOTES:
OBS= Financial Gift for David Houston?

Monday April 9th (Pipeline 5 to Mile Marker 83)
- Leave at (insert time here)
- (add name here) will take our gear back to Dave Houston's
- Robinson will be paddling with us.
- Richard will pick up our gear at David Houston's and bring it to us when he joins at New/Old 72 (?)
- Chris Reyes will join us at (insert time) from (insert place) to (insert place).
- Arrive at (insert campsite here) at (insert time here)

NOTES:
JOHN= Chris is joining on this day right?
RICHARD= Are you doing the above?

Tuesday April 10th (Mile Marker 83 to 65 a.k.a. Ghost River Lunch Stop Area)
- Leave at (insert time here)
- Arrive at Ghost River Lunch Stop at (insert time).

Wednesday April 11th (Mile Marker 65 to 48 a.k.a. Between Moscow & Rossville)
- Leave at (insert time here)
- Rachel will be joining us at (insert time) from (insert place) to (insert place)
- Have lunch at Wolf River Cafe?
- Pick up the Lagrange Mayor at (insert time) from (insert place) to (insert place)
- Arrive between Moscow & Rossville at (insert time here).

NOTES:
RACHEL: Are you joining us this day?
OBS= You two will need to work out who uses the SUP if you're both on the water at the same time.
OBS= You may want to ask Adam about the Outdoors Inc. SUP or Mark Babb about his.
JOHN= Will you have room if Rachel or Robinson need to join you?

Thursday April 12th (Mile Marker 48 to 31 a.k.a. South Of Collierville-Arlington Rd.)
- Leave at (insert time here)
- Arrive South of Collierville-Arlington Rd. at (insert time here)

Friday April 13th (Mile Marker 31 to 17 a.k.a. Between Germantown Pkwy & Walnut Grove)
- Leave at (insert time)

- 3pm arrive at Houston Levee to pick up High Schoolers
- Drop off High Schoolers at Germantown Pkwy. by sunset.
- Arrive somewhere between Germantown Pkwy. and Walnut Grove before dark.

Saturday April 14th (Final Day) (Mile Marker 17 to End)
- 7am meet volunteers at Walnut Grove.
- Meet Fathers and Sons with breakfast and coffee
- (insert time here) begin Father/Son Paddle
- (insert time here) Arrive at Kennedy Park and meet with College groups.
- (insert time here) Depart Kennedy Park
- (insert time here) Arrive at Greenway Park for final celebration!!!

Richard Sojourner replied:

RE: Question to Richard "Are you doing the above?" Answers: Helping with kickoff on Sat. YES; Bringing gear to HWY 72 Monday. YES; Joining in for the duration of the float Monday. YES. I can't make the meeting Friday evening at Dale!

Richard

John Henry replied:

I will be bringing Luke's board from James Lowery's to David Houston's on Sunday for Robinson. Christopher Reyes, and Phillip Beasley will be meeting me and Robinson @ David Houston's place on Sunday afternoon. we will all ride in together.

I don't feel like we will need logistical support (other than someone to camp with us and drive our gear from #5 - Robinson said he had a friend to do that) until after Monday evening. I feel sure that Richard will be able to bring any food, Ice, beer, necessary, when he comes.

As to compensating David houston - we can work that out @ Dale's Friday. Whatever I missed - let me know! :-)

As to Keith - richard and I were thinking that for him to drop his vehicle @ James Lowery's place on the way to Baker's Pond might be best... James Lowery's address needs to be forwarded to Keith, as I don't have it. Otherwise we could meet him @ the Citgo @ Hwy 5 and Hwy 72 @6:45...

Ray Graham wrote:

Yes, I want to drop in a couple times. I'd like to hike with you guys from Baker's Pond to the confluence with Goose Nest Creek to do a YouTube vid for the WRC. Then from Walnut Grove to Mud Island. I can help out with logistics either Saturday.

Ray

While the other SUPAS members were working out their last minute logistics, I was dealing with logistics of another kind.

I accepted BKK Transport's bid of $1270 and set the shipping wheels in motion. They quickly replied with a confirmation. Hoping to expedite things, I sent an email containing pertinent pick-up and drop-off information to a BKK dispatcher named Sherman.

Hi Sherman,

I booked a shipment with you through UShip today. I'd like to get pick up details so I can forward them to my contact in Eugene: which date you can pick up the bike in Eugene, what time of day, etc. My Eugene contact's name is Paul. Paul says 4:30-5:00 pm is best time of day for him for pick up. There are no street restrictions on North Polk Street in Eugene. You should be able to pull up in front of his house and load the bike. As you can see in the pictures, the bike is on wheels and will be easy to load on a truck with a ramp or a lift. Paul can assist you if needed. If possible, it would be great if you could take pictures of the Bikecar being loaded and secured on the truck. And, if possible, please email two or three of these pictures to this email address rod@**********.com. We will use the photos for promotion of the upcoming Bikecar journey from Memphis to Miami.

On the Memphis end, again, there are no street restrictions. Only foreseeable issue is that the house is 400 feet from the road. The Bikecar will need to be pushed up the driveway to the house. I can provide assistance if you want. I will need to arrange for someone to be home, so I will need an approximate arrival date and time. You can do this enroute if you feel it will be more accurate. Even still, I will not likely be in Memphis when the sipment arrives, so I will still need an approximate arrival date and time to pass on to my contacts. I will this need info today or tomorrow.

I look forward to hearing from you.

Cheers,

Rod

BKK replied:

Hi Rod

Thank you for accepting our bid. We will call you and we will call the pick up location to arrange the shipping your car bike shipping. We will call ina dvance as soon as we schedule the shipment. Looking next week sometimes hopefully. There is possibility to pick it up this week but we will update you either way.

Thanks
Best Regards
BKK transport & brokerage

And to make sure we were both on the same page, I shot off another email to BKK with full pick-up and drop-off addresses and phone contacts.

Greetings,

Not sure if you have pick up and drop off contact info so I will include it here. (N.B. Contact info omitted.)

Neither Dale nor I will be in Memphis next week. We will be on a river expedition and may not be able to be reached via cell phone, hence Meriam's cell number and the home phone number (land line). Meriam is Dale's wife. If you happen to know an approx delivery time by Friday night (Apr. 6) and can email that to me Friday night, that will help me arrange for someone to be to here to receive the bike. Dale and I leave for the river early Saturday morning (Apr. 7).

Feel free to call or email me if you need more information.

Cheers,

Rod

BKK Transport replied:

OK . We will do.

And with that, the work was pretty much over. Now it was up to Paul Adkins to do his part.

"Next week, while we're on the river," I said to myself with a comforting smile, "they'll call Paul, set up a time, drive by, pick up the bike, and two days later it'll be dropped off at Dale's. Just like clockwork. For now, my job is to just sit back and smile."

I sat back and smiled.

And then, BKK chimed in…

We want to pick up your bicycle today but we can not reach the person at pick up location . Please help us

Thanks

BKK transport

> "SHIT!" I shouted.
> I hurriedly pounded out an email to Paul Adkins...

Hi Paul,

Hope all is well with you.

A shipping company wants to pick up the bike today, Thursday April 5. I need you to call the shipper ASAP to arrange a time that works best for you. They have a truck in your area. BKK Transport - ***-***-****. If you are unable to be at your place today, we can reschedule something for early next week. However, it would be great to get the bike Memphis-bound ASAP.

Please send me a confirmation email so I know you have spoken with BKK.

Again, this is urgent. Both the shipper and I have tried unsuccessfully to contact you via phone. The number I have for you is ***-***-****. If you can supply us with an alternate number that would be great.

Cheers,

Rod

> I then sent an email to BKK Transport to tell them that I just sent an email to Paul Adkins.

Greetings,

I just sent Paul Adkins an email stating that he is to call you ASAP. Feel free to contact him via email as well. Sometimes it is easier to reach him via email rather than phone.

BKK Transport replied:

Ok. Thanks

BKK transport

And then, Paul Adkins chimed in…

Calling BKK now. No answer. I'm currently available at my office, at ***-***-****. Ask for Paul.

Paul Adkins

And then, BKK Transport chimed in with an email addressed to Paul and CCed to me…

Hi Please call us as soon as possible . Your customer hired us to pick up and ship the 4 wheel bicycle .We need to make arrangment to pick it up today. Thanks

BKK transport call us at ***-***-***

And then, I wrote another urgent email to Paul Adkins…

I talked with the dispatcher. He knows you are available to help today. The driver will contact you this afternoon. Sorry, don't know when he will call. They are keen to pick it up today.

An hour later, Paul Adkins chimed in…

Rod,

Called BKK and left a message (2nd time today) with my work phone. ***-***-****. Feel free to call to make the arrangements.

Paul

I immediately fired off an email to BKK Transport…

Paul Adkins in Eugene, OR - work# ***-***-****. I have also relayed this number to the BKK dispatcher over the phone.

Ten minutes later, Paul Adkins chimed in…

I just talked to Ian (dispatcher) - waiting for Driver to call me.

Paul

"Bloody hell!" I shouted. "When does it end???!!"
Two hours went by without a peep from anyone. And then, Paul Adkins chimed in…

Rod,

I spoke with driver - Mike. He is expected in town this afternoon - 4pm or so. I'll be meeting with him then. I'll give you an update once he has the bike.

Paul

"Thank Christ!" I shouted, relieved that some fucking progress had finally been made. I sat back (sans smile) and inhaled deeply. Then exhaled. Then inhaled again. Then exhaled again. Despite the deep breathing, the stress would not melt away.
Ten minutes later, Corn chimed in. He was about to leave Honolulu…

Mate, about to leave Honolulu. Let's lock in that bikecar courier if you can. We'll discuss money tmro but let's make it happen, these chances don't come along everyday.

See you in the morning

Dave

"Yeah, yeah, yeah," I said in my best New York bookie accent, waving my hand at the laptop like I was blowing off an annoying hustler, "I'm fackin' woikin' onnit!"

Two minutes later, BKK Transport chimed in…

We got hold of Paul already .Now we arranging the pick up of your 4 wheeler bicycle. We will update you later on

"Yeah, yeah, yeah," I repeated. "I fackin' knew dat awreddy."

I sat back and exhaled. A smile came over my face for the first time in hours. I chuckled.

"Looks like it's all falling together," I said.

And then, Corn chimed in…

Guys, bad news. I'm in Honolulu airport and they just bumped me off an overbooked flight. Seeking out alternative options now and will keep you posted as soon as I have info. Having not made it onto this flight I definitely will NOT be arriving in Memphis tomorrow morning. Doing everything I can to get there in time for Saturday morning.

Will let you know when I have updates...

Dave

"Well," I said, shaking my head in disappointment, "maybe I got cocky too soon. Crap!"

30 minutes later, Corn chimed in again…

A lot of outgoing flights are full here (who knew that everybody wanted to LEAVE Hawaii!) so I'm going to be on Standby on all flights until I can get onto one. Can you send me your cell numbers so I can send a quick message if I get on one...

If I can get out tonight it'll be on a plane to Salt Lake City then I'll have to wait at that end for an onward flight to Memphis, still smiling! Keep you posted...

Minutes later, Dale Sanders chimed in...

Dave,

Cheers Brother- Everything will work just fine. Congrats on your sailing journey. From one old sailer to a young one: "Faire Winds and Following Seas".

I chimed in with a dose of hopeful humour...

DC,

Memphis has a surplus of Nutella with your name on it! Looking forward to seeing you soon!

(N.B. Dave Cornthwaite has been trying to secure a sponsorship with Nutella for years. So far, his requests have gone unanswered. He says he refuses to give up.)

I decided to check in with Paul Adkins...

Hi Paul Adkins,

Checking in to see if you hooked up with the BKK driver.

Paul Adkins replied:

We talked at home, but he is due here anytime. He was concerned about space on his trailer.

I'll let you know here soon.

Paul

And then, just as I was about to celebrate the Bikecar's liberation from Eugene, Oregon, Paul Adkins chimed in…

Rod,

Mike was just here and he was not able to get it on his trailer. Apparently not enough room.

So we are moving it back inside the fence. He said he would call Ian and discuss.

So - we are in a holding pattern.

Paul

I didn't even bother with the inhale/exhale nonsense. I simply let the rage *rage*.
"*FUUUUUCCCCCKKKKK!!!!!!!!!*"
An hour later, Jonathan Brown finally got wind of Corn's dilemma and chimed in…

Wow! I just popped into an email frenzy! Big bummer you got bumped. don't feel dumped but instead get pumped cuz you're about to come back to Memphis where you can get jumped!

Lemme know what the haps is.

JB

The last words of the day were Dale's. His message was a grounding, of sorts—a way of saying that tomorrow is another day, another chance to get things right. Optimism, adventure, and full plates of food lay on the horizon. With this kind of Southern hospitality, the ugliness of the day didn't seem so bad after all.

"Now, if we can just get Dale to remember what day it is," I said to myself, chuckling, "we'll be doing even better."

Tomorrow evening, 6:00 PM, Friday 7th (N.B. Actually, it was the 6th.), *Meriam will have food for all. Don't worry about bringing anything other than an empty stomach. We will load craft and supplies and then eat, as well as lots of social good conversation —- last minute trip details. VR Dale*

Shuttle logistics were the main topic on the final day before the SUPAS launch. Getting six paddlers and their watercraft to Baker's Pond was a task in itself. Keep in mind that these paddlers were heading one-way without gear, so their vehicles had to be returned to Memphis while their camping supplies and food needed to be dropped at the location of the first camp—in this case a riverside property owned by James Lowry. On the morning of Day 2, the camping gear and food would then need to be shuttled to the location of the second camp. The same for Day 3. And because the number of paddlers varied for each day (six on Day 1, three on Day 2, seven on Day 3), the extra watercraft (canoes, kayaks, and SUPs) would need to be shuttled to each camp location and either dropped off, or hauled away. Add in the fact that the Day 2 camp location was in a very remote setting, accessible only by 4x4, and you've got a huge logistical puzzle on your hands. If not carefully planned out, it could easily, and quickly, turn into a nightmare. The time spent planning the shuttle logistics for the first three days of the expedition was mind-numbing. Dale, JB, John, Mike, Rachel, Ray, Richard, and Robinson had done an amazing job of sorting out the details. And it needs to be added that this was an *eight*-day expedition. There were five *more* days of logistics to follow. Super-duper kudos to all involved!

On April 6, 2012 at 1:27PM, Dale Sanders wrote:

I have arranged Boards & Boat on my trailer —- all fit well. Will be using Van vice Truck, which means I can carry all paddlers except John and Keith. John will be able to drive straight from his home, saving mucho time. Keith will be driving his own vehicle; he will only be paddling the first day. If any of you live close to John, riding out with him is still and option. "Paddler Choice", so to speak. Will see you tonight for the big gear loading and trip kick-off. Food tonight here - OK? Realize some of you can't make it but thought I would keep you informed. Mike might give us a big surprise - on Thursday we need to keep an eye to the rear, once in a while. Ray' are you available anytime during the week to help at access points like Bateman Road Etc.? High probability we will need someone to shuttle the Mayors back to their vehicles at Leatherwood. Dale

…and what about ol' Corn? Who was gonna deal with the missing Brit?

On April 6, 2012 at 5:01PM, Dale Sanders wrote:

As you know by now Dave was bumped off his flight yesterday, which would have put him in Memphis (MIA) at 0730 AM earlier today. Dave now has tried some 20 flights - all full going out. Just spoke with him on the phone few minutes ago. Best he has confirmed at present is a Honolulu departure tomorrow for LAX. At present, he has no confirmed flights to MIA but feels he will be able to work something out once he arrives in Los Angeles. Looks like his arrival time MIA will likely be sometime Easter Sunday. He plans on joining us when he arrives. Most likely he will not arrive in time to paddle with us Sunday. The logistic could be difficult, but, it is possible Dave could ride out on the Four Wheeler Sunday afternoon to the gas line #5 Campsite, spending Sunday Night there with us, paddling Monday through the SUPUT Team folks declared, most beautiful swamp on the Wolf. The other option is meet us at New Hwy. 72 Monday afternoon with Richard.

If you, or anyone you know, have any idea(s) how Dave could be picked up at the MIA, helped with packing and taken to the Rivers edge please let me know ASAP. I will not be able to coordinate any of these logistics after early AM tomorrow. Any suggestions please call - Cell: ***-***-**** or Email. Any help, at this stage, will be appreciated.

Rachel Sumner replied:

I'm sure I could get him once we hear something... And get him out to the fourth wheeler... please, let me know mite details as they arise.

Could you let him know to be in touch with me? He should have all of my contact info from the last visit.

Thanks for working on this, Dale!

Rachel

Dale Sanders replied:

Rachel, You and Robinson were God sent. Wonderful if you and Robinson can get him to the pipeline for camping at #5 Sunday night. If by some chance he doesn't arrive Sunday in time could you help getting him to the rivers edge Monday afternoon? Richard may also help for he will be joining us then at New Hwy. 72. Mike plans to start paddling Tuesday (time unknown) at New Hwy. 72, fast to catch us by Thursday (or possibly Wednesday night). If by some chance, Dave doesn't arrive until too late Monday to paddle he could always start his paddle with Mike Tuesday.

On April 6, 2012 at 5:02PM, Rod Wellington wrote:

Hey Dave,

I was talking to John Henry regarding snake protection. He has offered to lead the team through the swamp, which means that he will deal firsthand with the snakes. "Of course," he said,"it's usually the guy that goes second that ends up gettin' bit." That's reassuring. :)

Dale and I went to a local hunting shop yesterday and looked at snake protection options. I can't justify spending $60-$100 on specific gear for snakes when I will only be using it for three days and then not use it again.

Any update from your end?

Rod

Dave Cornthwaite replied:

Hi Rod,

I hear you on snake protection. Sadly, looks like I'm going to miss that section. Updated Dale earlier, would love to get all of you guys on a skype call on Dale's big screen tonight. Want to talk about filming and what I can do even when not paddling, as well as check in and wish you all a good start....

On April 6, 2012 at 8:06PM, Dave Cornthwaite wrote:

Hello all,

It would appear that I'm now extremely familiar with Honolulu airport. Easter weekend was not the weekend to fly out, it seems.

As you'll all be together tonight at Dale's I was hoping we could have a group Skype. As well as updating you on my arrival I'd like to offer some filming tips and advice now I won't be there for the first couple of days, and also record your thoughts and feelings so I can write a blog tomorrow to keep up the great momentum that you've all created so far.

It'd be super to find out what media attention has been generated to date,

I have some contacts from last year's Mississippi paddle and will do my utmost to ensure some extra publicity for the end of the trip.

I am so disappointed not to be there tonight. As much as I abhor wasting time the upset of not being able to join the entirety of the Wolf River Paddle far outweighs any innate dislike I have for airport food! In order to jump past the standby nightmare I've bought another ticket for tomorrow, but the staff at Honolulu airport (we're now good friends) say I'd be lucky to get my onwards flight to Memphis before Monday morning as there's only one flight out on Sunday, and again I'll be on Standby. So, I expect to see you guys on the river Monday evening. Please don't make any jokes about English people avoiding hardship!

Hope we get to speak tonight. I can be online anytime, let me know a time if Skype is possible (I see my big face on Dale's plasma screen, once more!)

Until then...

Dave

Memphian Boyd Wade, an avid paddler who had built a close friendship with Corn during his 2011 Mississippi descent, replied:

Dave,

Sorry to hear of your poor flying experience. I hope your new itinerary stands.

I can be available Monday afternoon with or without 4 wheeler to help get you on the river. I do have a 6:30pm commitment that evening.

Best of luck. And let me know how I can help.

Boyd

The food that night was amazing, with plenty of vegan options for yours truly. Meriam Sanders (Dale's wife) is an incredible hostess. Her culinary skills are off the chart. The SUPAS team were a stuffed and happy lot by meal's end.

Our Skype chat with Corn was typically amusing, despite his unfortunate circumstance. Unable to leave the Honolulu airport (in case an opening came up on a Memphis-bound flight), Corn was making the most of his time, catching up on business-related loose ends and, as always, planning future adventures.

Final SUPAS logistics were hashed out among the team members. Paddleboards and canoes were secured to the trailer. Weather forecasts were checked. Alarm clocks were set. Heads met pillows, and the final night of comfy sleep ensued.

Jonathan Brown—ever the busy boy—chose not to bed-down. He'd left his gear purchases and packing to the last minute, and was now surrounded by a heap of shiny new camping gear, electronic devices, clothing, and discarded packaging. Aquapac, a British manufacturer of waterproof bags, had signed on as a SUPAS sponsor (thanks to Corn's involvement), and Jonathan was busy stuffing cooking pots, cutlery, fuel canisters, a Jetboil stove, packages of dehydrated food, a tent, and a large sleeping bag into a 35-litre, orange, black, and blue Aquapac bag.

Content with his efficient packing skills, JB turned his attention to a small, bullet-shaped camera. Its gaudy colour (canary yellow) lessened the likelihood of it being misplaced in the swamp or amongst his gear. He carefully affixed the camera to a mount and attached it to a wearable headband. He slid the headband over his shaved scalp and checked his reflection in a mirror.

"Looks manly," thought JB, turning his head to see every sexy angle of his bald palette. "I'll get all the SUPin' action, fo' sho!"

He removed the headband and stuffed it in his pack.

"I couldn't afford a GoPro," he'd told me earlier that night, "so I got *this*. The audio's pretty crappy, but the footage isn't too bad. It looks hi-tech in a Flash Gordon kinda way, doesn't it?"

I smiled at his gentle sarcasm and nodded my approval. JB had a knack for finding a silver lining in anything.

Thanks to his employee discount at Apple (where he worked one-on-one with new customers), he'd purchased an iPad for the expedition and loaded it with the detailed maps SUPAS member Mike Watson had compiled. When linked to the iPad's GPS, we would be able to view our position on the Wolf and cross-reference it with the routes that the earlier scouting missions had taken. Thanks to their trial and error attempts at finding a navigable course through the swamps, we hoped to avoid the trouble spots and move downstream relatively unhindered. Our plan, of course, sounded far more convincing in the comfort of Dale's basement. The real test would begin in the morning.

JB finished packing his gear and glanced at his watch as he shut off the overhead light. It was 4:00am. The wake-up call would come in two hours. He knew sleep would elude him tonight.

He laid back on the couch and shoved a well-used pillow under his head. The charging lights from his plugged-in electronics cast a dim red glow over the darkened room. He closed his eyes and thought back to a time far removed from the present. It was a time he'd tried to forget, but its adventurous relevancy had ensured its staying power. In the morning, he and two friends would attempt the first continuous descent of the Wolf River from its source to its mouth with watercraft in tow. Another friend, John Henry, had accomplished the feat in sections. Others, like Gary Bridgman and Bill FitzGerald had walked the first 12 miles of the river before beginning the paddling portion of their journey. Unknown to many was the fact that JB had tried his hand at descending the Wolf eight years prior. Tomorrow, he would begin his *second* attempt. It was the very thought of that first disastrous effort that kept him awake on this night.

On July 10, 2004, JB and his friend, Kevin Brunson, set out from Chapman Road Bridge in a canoe with a naïve belief that they would paddle all the way to Memphis. That plan went sideways from the get-go. They struggled through dense buckbrush and sharp sawgrass, dragging their canoe more than paddling it. Bruised, bloodied, and downright exhausted, they left their boat at the river's edge and stumbled through thick forest to a rundown house trailer, hoping someone would be home. A man answered the door and took them in. He offered them food and a shower, and provided dressings for their wounds. That man—*their* River Angel—became known as Trailer Jon.

Despite the tortuous events of that ill-fated day, a day that saw them travel less than three miles in 12 hours, a burning desire to have another go at descending the Wolf gnawed at JB. *This* time, instead of being encumbered with a weighty canoe, he'd use a much lighter craft: a stand-up paddleboard. And he'd bring along some friends, *a lot* of friends; hearty, adventurous friends well-versed in arduous river travel.

JB pitched his idea to his new friend, Dave Cornthwaite, back in August 2011, when JB, Dave, Dale Sanders, Dave's brother Andy, and myself paddled down the Mississippi River from Memphis to Tunica. Dave loved the idea and assured he would partake.

Realizing the opportunity to do the trip as a fundraiser, JB contacted his good friend Rachel Sumner, who, as well as being a volunteer with Operation Broken Silence, had also been instrumental in organizing the media exposure for Dave's arrival and departure in Memphis in 2011. Together, JB and Rachel laid the groundwork for SUPAS.

In a few short hours, JB would get another crack at descending the Wolf. This attempt would be thought out far beyond the disaster that transpired in July of 2004. With the help of many River Angels, JB hoped to arrive unscathed and smiling at the same property where he'd met Trailer Jon all those years ago. And then, instead of going home defeated, he would move downstream emboldened. This attempt would be different. This attempt would be legitimate. This attempt would be *successful*. There was one thing, however, that stood between him and success. That *thing* was a 102-mile-long obstacle course called the *Wolf River*.

SUPAS – Day 4 – April 10, 2012

Once the bitterness took hold, it was impossible to shake it. It embedded itself in my brain and grew out of control like a late-detected cancer. No amount of positivity could deter it. No amount of comradery could cure it. Its grip was tight and lethal. It cared about nothing but the annihilation of its host. It was a silent killer. It was mine, *my* bitterness, but it was using me to destroy me, to destroy those around me, to sabotage the good in my life and replace it with hate. It made no apologies and adhered to no morals. It screamed in silent fury at anything that opposed it. It was here for the long haul, here to the end.

I fed it with paranoia and blind rage. It ate my fears and vomited derision. Its presence ensured complete destruction of any bonds built between myself and others. Its only goal was total isolation, the absolute riddance of all but itself. And when its goal was finally reached, it would backpedal and craftily rein in those it had callously alienated, thus beginning the cycle again.

Day 4 brought a new enemy into the SUPAS fold, a new target for my entrenched indignation. This new foe would join Dale Sanders and John Henry on my silent shit list. I would find ways to hate them without direct confrontation, ways that would prolong the animosity, ways that would keep them at bay, ways that would keep me safe in my fucked up comfort zone.

Dave Cornthwaite arrived at the end of Day 3 spouting cheery optimism and offering a fresh perspective on the drama currently unfolding on the Wolf River. At first, I was relieved to see him. His presence allowed me to vent at length about how Dale had created an awkward situation earlier that day by threatening to leave the team behind and forge on alone, and then actually following through on his threat by abandoning us at day's end. Dave listened intently to my angry rant and responded thoughtfully with soothing words of support. I went to bed angry that night, but not as angry as I would've been without his help.

Morning brought a new, unexpected challenge that instantly turned me against Dave. It changed him from friend to enemy in a matter of minutes. What instigated the shift? A GoPro camera and Dave's incessant use of it.

The SUPAS team knew that Dave planned to create a short documentary about the expedition. He had made that plan clear early on. But that plan changed when he got stuck at the Honolulu Airport and had to miss the first three days of the expedition. To compensate for our absent filmmaker, Jonathan and I made every effort to film as much of the descent as possible. Jonathan's head-mounted camera was easily accessible. For him, filming on the fly was not an issue. My camcorder, however, was stowed in a waterproof drybag strapped to the deck of my SUP. Each time I wanted to film something, I had to stop paddling, unstrap the drybag, retrieve the camera, film, return the camera to the drybag, and strap the bag back on the deck. It was a tedious and time-consuming process, but I felt it was worth the effort. I was documenting the expedition. I was contributing to the creative aspect of the project. Doing so helped relieve some of the stress I was accruing from the physical toil I was experiencing. Filming became a welcomed distraction from the brutal reality of trudging through a snake-infested swamp with people I didn't really like. So I filmed. A lot. Doing so ate up time. The more time I spent filming, the more annoyed Dale and John became. I was holding them back. The team wasn't progressing as fast as they wanted and it was *my* fault. I fucking resented them deeply for the guilt I felt. Had I been on a large river with no obstructions (like the Mississippi River, for example), I would've told them to go ahead without me. But we were moving slowly through a god-forsaken swamp—a swamp I knew nothing about. I needed them more than they needed me. They knew that. They complained daily about my dallying, but they never abandoned me. They always waited around a bend for me to catch up. I'm thankful for that. But I'm not thankful for the incessant guilt I felt.

When Dave joined the fray at the end of Day 3, he assumed the role of chief documenter. He pointed his GoPro at everything. Unlike my camcorder, his camera was always at the ready which meant he was filming far more than me, much to my displeasure. Very reluctantly, I handed the filming reins to him. I wasn't going to compete with him. It was impossible. He was far more familiar with filming than me. He knew when and where to point the camera. He knew the importance of cut-aways. He interviewed us. He captured our struggles and our triumphs. And he laughed. Often. Too often for my liking.

I had been harbouring a huge amount of hate for three long days and that anger was bound to surface somewhere. I wasn't happy. Not even close. I wasn't smiling or laughing. I was secretly wishing that my teammates would fuck off and leave me alone. The one thing I clung to, the one thing that was keeping me sane (the filming), had been taken away from me. More correctly, I had *given* it away, and I was furious with myself for having done so.

I sunk into isolation and built an oversized grudge against Dave. I was jealous when he and JB paired up and paddled side by side. I was the odd man out. I drifted and fumed and brought up the rear. I kept everyone at a distance, rarely smiling, rarely talking, rarely caring whether we made substantial headway or not. Dale tried to engage me in decision-making—which direction to go, when to call it a day, where to pitch camp—but I paid no attention to him and only offered shrugged shoulders and averted eyes. One hour blended into the next without my noticing. My surroundings, my teammates, my love of being outdoors all faded to nothingness. Details became distant. Location and safety became irrelevant. I quietly cursed when obstacles appeared. I frowned when the channel split into four and I had to choose a new course. I shut myself away inside my tent at the end of each day and refused to interact with the others. They were unwelcomed company. I silently competed against them in a one-sided game—a game of which only I knew the rules. If they wondered whether I might crack and quit, I'd prove them wrong and continue on. I had given them power, but I would not give them conquest. I would be defiant to the end.

During one instance where I actually paddled and conversed with JB and Dave, JB quizzed me on aspects of my past expeditions. Satisfied with my answers, he asked,

"Do you think you'd ever go on an expedition with Dave and me?"

My reply was quick, sharp, and final,

"Nope."

The team was joined on Day 7 by a family of three, the only participants in the Father and Son Paddle portion of the trip. Its intentions were to get people involved in the journey, to promote awareness for Operation Broken Silence, and to generate donations. We were expecting more people to turn out for the event, and were disappointed by the lack of enthusiasm. Regardless, the father and mother paddled a canoe donated by Ghost River Rentals (a local outfitter) while the eight-year-old son joined JB on his SUP. The family seemed to enjoy themselves during the hour they were on the water. I kept quiet and observed.

JB seemed perfectly adopted the role of "big brother" when paddling with the young boy. He showed the boy how to hold the long SUP paddle and how to propel the craft in a straight line. They bonded like old friends and I found myself feeling jealous about their connection. JB had a knack for engaging others. He had outstanding people skills. I did not. That fact generated more jealousy, envy, and anger toward JB. These aggressive feelings had been precipitated by a curt suggestion JB made to Dave and me while the three of us talked during a break prior to meeting the aforementioned family.

My memory of the conversation and its contents are hazy at best, but I remember clearly that the tone was dark. We discussed depression, hardship, abuse, and my lack of interest in having to deal with a family of strangers joining us for an hour-long paddle. Being part of the team meant I would need to interact with these strangers. I saw that as an unnecessary challenge, something that would just add more stress to an already stressful situation. JB countered with his secret weapon: positivity. He reassured me that all would be fine and that I would enjoy the family's company and that their presence would enhance our trip. I rebuffed his pep talk. Surely, all those years as a Christian missionary had helped shaped his drivel of unpalatable positivity. As the three of us mounted our boards and shoved off from the riverbank, JB's curt advice stuck in my head,

"Let's put on our *people* faces and go meet the participants."

"*People* faces," I thought to myself, repeating JB's words with a high dose of disdain. "Sure, for *you* I'll paddle into this unknown situation with a fake smile and a false care. I don't mind doing that for you, JB. All for you and your vomit-inducing positivity. I'll create a façade so that Operation Broken Silence and SUPAS' reputations won't be harmed by errant comments from some anonymous family who disliked the company of some grumpy, dreadlocked Canadian. I'll do that for you and your *fucking* charity event! I'll do that to keep the drama and fisticuffs at bay. I'll do that like every good Christian would do—all those fake-happy fucking church-goers you see every Sunday morning! Just for you, just to keep the peace, I'll be what *you* want me to be. I'll appease *your* feelings of contentment. Never mind that it'll be fake and fucking transparent! Never mind that I'll fucking hate you for it for years to come! Never mind that I've been tolerating you and the rest of these fucking bozos I've had to paddle with for the past six days! Never mind the snakes and the swamp and the intolerable stress I've been under! Never mind *all* that fucking shit!!! I'll put all that aside and do what will make *you* happy: I'll *smile*. And when I do, please know that if I had a knife in my hand, I'd bury it deep in your neck and leave you floating facedown in the river. JB, FUCK YOU and your *people* faces! FUCK YOU and your fucked up Christian *values*!! And while I'm at it: FUCK CORNTHWAITE and his goddamned GoPro camera!! FUCK him and his leadership skills!! FUCK him for doing no wrong!! FUCK him for being everyone's friend!! And FUCK him for flying all the way from Hawaii to participate in this fucked up charity event!! FUCK YOU, FUCK DAVE, FUCK DALE, AND FUCK *SUPAS*!!! THE LOT OF YOU CAN GO STRAIGHT TO *FUCKING* HELL!!!"

Streams of hate like the one above surfaced dozens of times per day during my time on the Wolf. My time on the river was not a happy one. I came in with expectations of broadening the public's awareness of me as an "adventurer." I hoped that SUPAS would help me reach the next level of success in this unorthodox career I'd chosen. I clung to that hope all through the planning process. Back then, when I was receiving scads of emails every day, I felt *part* of something. I was part of a group, a group of like-minded people coming together with good intentions. I enjoyed that sense of comradery. My contributions were welcomed whole-heartedly. But when the planning phase ended and the action phase began, everything changed. To me, that fucked up transition came as no surprise. It happens every time I embark on an expedition.

I have many masks that I wear. Sometimes I'll show the world a smile. Sometimes I'll put on a face of indifference. Sometimes (mostly)

I don't show my face at all. I lock myself away behind closed doors for fear of what society thinks about me. (Sometimes, I don't give a fuck *what* society thinks of me!) I'm a rebel, a loner, and a really great guy. I'll help you. I'll hurt you. I'll leave you to die. I can be polite and trustworthy. I can be corrupt and deceitful. I can shake your foundation or completely ignore you. Simply put, I am not what I seem.

When it comes to expeditions, I possess two personalities: the planner and the doer. The *planner* arranges all the pre-expedition logistics and assures that things will proceed as planned. He buys the maps, makes the phone calls, writes the sponsorship proposals, spends hundreds of hours researching data online, frets about *everything*, and does his best to promote the fuck out of an upcoming expedition.

The *doer*, on the other hand, is assigned with the task of performing that which the planner has planned. Typically, the doer resents the planner for creating such a huge fucking workload. The planner pays no mind to the doer because the planner doesn't have to perform the task. I compare it to a large business environment where the administrative staff create and implement work for the labourers. The labourers resent the admin staff for lumping a large workload on them, but inherently know that their job is to follow the orders sent down from higher up. The labourers understand and resent the hierarchy even though they take part in the game. Their attitude is self-defeating. They alone must claim responsibility for their role in the hierarchy. Most of them are obedient slaves. They follow the orders handed to them and stave off their desire to complain. They may grumble, but they rarely revolt. Those that revolt typically become aware of the game, voice their displeasure, and, at some point, quit their jobs and go off in search of greener pastures. Doers, it seems, are never happy people.

Planners, on the other hand, are not happy people either. Their jobs come laden with responsibility and they, too, have to answer to higher ups. No matter the position, there's always a higher up. It's one of those fucked up burdens of being human—the constant reminder that someone or some*thing* is more important than you. I have come to learn that people or things are only as important as I make them. *I* assign importance, not the other way around. I can enter into a situation with a police officer and know that the police officer has only as much authority over me as I wish to give him or her. In reality, that police officer exists on exactly the same level as me. He is my peer, and nothing more. His "authority" over me is imagined. The "authority" exists only if I comply with the notion that it actually exists which, of course, it does not. If I believe the officer has no authority over me, then he has no authority over me. If a dentist tells me she is a dentist and I don't believe her, then she is not a dentist. If a crazy person on the street tells me they are the president of the United States, I won't believe them. *They* may believe they are, but I *know* they are not—I don't buy into their game. People will have us believe that they are any number of false personalities. The idolatry of celebrities is a fine example of this.

Celebrities are nothing more than admin staff who saw through the bullshit game and found successful ways of promoting themselves as marketable brands. Do you have big tits and a pretty face? Congratulations, you're a fucking celebrity! Do you have a long cock and bulging biceps? Congratulations, you get to fuck the other celebrity and become a celebrity yourself! Isn't life great!? All you need to do is believe in yourself (and dupe everyone around you into believing you're fucking top shelf material) and presto! ... you're waving goodbye to the admin staff *and* the labourers. Of course, you better keep in mind that your newly self-created workload will be weighted down with heaps more responsibility, assumptions, expectations, and disappointments. Like it or not, you'll still have to suck a lot of cock and you'll still have to take it up the ass on a regular basis.

Life is all about getting off. It's how life starts, it's how life continues, and it's how life ends. Life is a bareback fucking carousel ride from hell. We get on. We get off. We fuck. We get fucked. We come. We go. We put someone in charge because we bought into their game. That person bent us over and fucked us hard. It sucked. We believed their lie and paid the price. Hierarchy exists because the drones believe it works for them. At the same time, they know it's a farce. They know it's a game, but they continue to play. Why? Because it's safe. It's comfortable. All you need to do is breathe, relax, and take it all in. The dong fits because all the others that came before stretched you out good.

And while we're on the topic of grandiose butt fucking, this would be a good time to slide in a Kanye West quote. Here's a lyric from his song, *New Slaves*: "You see, it's leaders and it's followers, but I'd rather be a dick than a swallower."

Take away what you want from that quote. In the end, the choice is yours: Lead or follow? Fuck or swallow? Truth is, we tend to do all four, on all fours. And in doing so, we fuck away our future.

So what does all this have to do with the planner and doer personalities I alluded to earlier? Not much, really. It's just a roundabout way of saying there exists a rift within me that I'm well aware of, and would rather maintain the chasm rather than bridge it. If you think about it, it's really the basis of this entire book. The planner enjoys the company of others (from afar), but detests their company face to face. It's all about contradiction. It's all about simple yet complex paradoxes. It's all about creating a persona and tricking everyone into believing you're the next big thing. It's all about truth and illusion, reality and delusion. It's all about maintaining the dichotomy. The planner compliments the doer—they are both in the same. Each knows the other exists, but they refuse to acknowledge each other in those precious moments of productivity. Each bends over for the other. Each acknowledges the hierarchy and each resents the hierarchy. The doer possesses more resentment because he comes second. His place in line is a result of the inherent hierarchy. Despite the doer's best efforts, he can do nothing to change the hierarchy. He must recognize it and accept it. But he is a rebel. He is defiant. He believes himself to be the dominant celebrity in his duel with the planner. He sees only the façade that the planner has manifested. The doer is blinded by illusion, deluded by inner trickery. When the time comes for the planner to step aside and the doer to perform, the doer is resentful of the work orders that have been handed down by the planner. At this point, the doer has two choices: accept or reject the orders. If he rejects, no expedition transpires and the planner's efforts are all for naught, which leaves the planner resentful. If the doer accepts, he must follow through with the orders, no matter the cost—and the cost is often high.

The transition time between the planning stage and the doing stage is the most critical, and it's usually where resentment and drama resides. This is the only time when both planner and doer meet face to face. It is the time when the baton is passed. If one of them fumbles, the race is over. It often takes considerable effort on the doer's part to assume his position. Up until this point, he has resided in his comfort zone. It is now his job to leave that place of sanctity and venture into the unknown. He carries a hefty manual, supplied by the planner, but the manual comes with no guarantees. The more the doer struggles with his concept of hierarchy and the enormity of the task ahead, the more the expedition is in jeopardy. The doer is a creature of comfort. He resists change. The more familiar life is, the more comfort he'll enjoy. Adventure holds little comforts. It presents only change, and change is not good in the eyes of the doer. He will feel the prodding of the planner during the transition stage. The planner *needs* to have his orders filled. There are deadlines in place that need to be met. But the doer is a rebel. He rejects deadlines. He prefers *life lines*. He prefers a safety net, yet he knows no safety net exists in adventure. Adventure is hardship and challenge, suffering and stress, exhaustion without reward. Basically, adventure *sucks*. But he is drawn to it, like a moth to a flame. Adventure will consume him. It will annihilate him. It will transform him. It will make him into a *planner*.

SUPAS – Day 2 – April 8, 2012

"Rod!" shouted Dale. "How's it look ova thare? See any water?"

"No water!" I yelled back. "Only a big meadow!"

I was standing on the edge of huge, unfenced open area bordered on three sides by dense forest. Water seemed to be draining from the meadow, but no surface water was visible. The flow was underground. About 200 yards to my right was a gap in the trees. I contemplated walking down to see if the gap allowed access to the river, but then thought better of it. I'd been dragging my stand-up paddleboard over earth and obstacles for a good part of the day and didn't need to create more work for myself. There would surely be more challenges downstream.

"There has to be an easier way through this mess," I mumbled as I stumbled down a muddy rivulet that branched and severed the swamp into a chaotic jumble of dead grass, fallen trees, and barely flowing water.

Our team of three had split up in an effort to find a navigable channel. Both Jonathan and I had left our SUP boards in a central spot on level ground and fanned out on foot north and south to search for a path through the chaos. Dale continued east, pulling his canoe behind him.

None of us had been in this section before. The only man on our team who had—John Henry—was shuttling our camping gear and food to Pipeline #5, the spot where we planned to meet him, Robinson Littrell, Chris Reyes, and Phillip Beasley at day's end. Somehow, I'd missed the memo stating John's absence on this day. I was looking forward to him leading us through the swamp and was disappointed, and frankly, a little scared, when he told me he wouldn't be joining us. There was no doubt in my mind that the three of us couldn't cover the six miles from James Lowry's house to Pipeline #5 without John, but having someone who had paddled this section before would've alleviated some of the endless guesswork that encumbered us on Day 2.

"I got water ova here!" shouted Dale. "Ware yoo at Rod?!"

"Not too far away," I said, just loud enough for the others to hear me.

"Jonathan!" shouted Dale. "Yoo steel with us?"

"Yes, sir!" hollered JB. "Right behind ya!"

Half-knackered and muddy up to my knees, I made my way back to my board, grabbed the nylon rope attached to the bow, wrapped it around my wrist, and set off in the direction of Dale's voice.

Paddle. Drag. Paddle. Drag. Paddle. Drag. Paddle. Drag. On and off the board a hundred friggin' times, until it becomes almost unbearable. And then…a long, wide channel canopied with early spring foliage. Beautiful. Paddle. Paddle. Paddle. Paddle. Paddle until the unenviable inevitable arrives. And then…Drag. Drag. Drag. Drag. Drag.

It wasn't uncommon to find ourselves trudging through dense forest, the canoe and paddleboards bonking off tree trunks and riding up over fallen trees. Bushes and low branches scratched our faces. We grunted, groaned, cursed our luck, sighed, and moved on.

Numerous times we consulted our GPS and satellite images and fanned out in search of even a *shallow* channel; anything we could float our crafts in, anything to ease the burden. Sometimes we found ankle-deep streams or long, stagnant pools. Many times we didn't. When we didn't, we would reconvene, consult the maps again, and use Dale's strategy.

Dale had a compass mounted on the bow of his canoe. When a direction had been decided, Dale would turn the boat's bow so the compass needle matched the chosen direction and proceed to drag the boat through the forest until a new direction was needed. We repeated this routine for what seemed like hours.

After much trepidation, we arrived at Pipeline #5 and were greeted by John, Robinson, Chris, and Phillip. Two full days of difficult swamp slogging had brought us 12 miles. I took little comfort in the fact that we still had six days and 90 miles to go. John and Dale assured me the worst was over, and that after the Highway 72 bridge, things would be easier. I took little comfort in their assumption.

"I'll believe it when I see it," I said to myself with a disparaging frown.

Thankfully, no major personality clashes occurred during Day 2. Perhaps John's absence had something to do with it. There was certainly a detectable unease in the air when John and Dale were around each other. Dale assumed leadership on Day 2 and he seemed to enjoy the responsibility and the challenge it provided. Early in the planning stage, he'd confided in me about his displeasure at Jonathan's lack of leadership. To his credit, JB saw no need to be a leader or assign a leader. He approached the subject democratically and encouraged others to step forward and assume temporary leadership. Dale, a disciplined Navy man, saw things differently. He liked leading, but seemed to abhor being led. Military is heavily based on giving orders or taking orders. You do one or the other, but rarely both. As a river guide with Wolf River Conservancy, Dale gave the orders, so to speak. He made the decisions and led the paddling groups. No one questioned his motives because he was usually the one with the most paddling experience. But when John Henry entered the SUPAS fray, everything changed. John had no interest in being bullied by river guides with military backgrounds. He had his own agenda, his own way of approaching the Wolf. With two headstrong paddlers competing for leadership of SUPAS, a confrontation was bound to happen. Surprisingly, Dale helped diffuse the situation by suggesting the idea of daily leaders. Following his own lead, Dale assumed leadership for Day 2. John led for Days 1 and 3—a wise decision based on his knowledge of the upper Wolf. Days 4-8 remained open to whoever wanted to lead. Not surprisingly, Dale assumed the role. John provided shuttle service and stayed mostly in the background during the second half of the expedition. He did, however, join us on Day 8 for the final 11 miles to the confluence of the Mississippi River. His interaction with Dale was minimal, but it was good to see him "wet a paddle" with us again.

John wasn't the only one joining us on the river on Day 8. His paddling partner, Ray Graham—who accompanied John during the early scouting missions—camped with us at the end of Day 7 and paddled the last 19 miles to the Wolf's mouth. It's worth noting that Ray is one of only four people who have paddled the entire length of the Wolf in sections.

Tom Roehm was there in his small motorboat. Tom supplied Dave Cornthwaite (and Dave's brother, Andy) with accommodations during their time in Memphis in August 2011. He'd also joined the SUPAS team on Day 5 when we paddled through the amazing Ghost River section, a flooded cypress forest that many people consider to be the prettiest part of the Wolf River.

Mapmaker and GPS wizard, Mike Watson, was there as well. Mike, whose detailed maps helped us successfully navigate through the treacherous upper reaches of the river, joined for the last five days of the journey and had been involved with the early scouting missions. Like Ray, Mike also paddled the length of the Wolf in sections.

Richard Sojourner joined us on Day 3 at the Highway 72 bridge and paddled the remaining five days to the river's end. In doing so, he earned a place alongside John, Mike, and Ray as one of only four people to canoe the entire Wolf in sections—an amazing accomplishment for a 67-year-old man who had undergone major heart surgery only seven years prior. His doctor told him he would never paddle again. Richard proved his doctor wrong.

Although he missed the first three days of the expedition due to a flight mix-up at the Honolulu airport, Dave Cornthwaite paddled the final five days with us, filming and smiling incessantly. I distanced myself from Dave early in the expedition. I felt he'd usurped (unintentionally, of course) my "celebrity adventurer" status in SUPAS. I resented his presence throughout, but also felt akin to him in terms of previous expedition experience. We'd both crossed the Australian continent under our own power (him on a skateboard and me on a bicycle) and we'd both paddled the Murray River (Australia's longest) from source to sea. I'd helped him with logistical support during the planning stage of his Bikecar expedition and planned to join him on the bike for a few days once the SUPAS journey ended. I wasn't sure how spending hours pedalling with him would play out, but I was willing to set aside some of my animosity and give it go. The impact of that decision comes later in this book.

Jonathan Brown and I became the first people to descend the entire Wolf River on stand-up paddleboards, a feat no one had ever attempted. We weren't presented with medals or trophies at the journey's end, but a large crowd consisting of Operation Broken Silence staffers, local media, followers of the journey, and members of my family (my father, Robert; sister, Carrie; and brother-in-law, Mario) provided all of the paddlers with a raucous reception that bested any shiny token of accomplishment. Barbara Blue, the Reigning Queen of Beale Street, and her gifted keyboardist, Nat Kerr, treated us to some tasty Delta blues. An Elvis Presley impersonator shook his hips and belted out some classic cuts by "The King." And local folksters, Star and Micey, rocked the crowd with an impressive set of cerebral singalongs. Huge kudos to John Henry for helping secure the city permits for Mud Island (the site of the party, where the Wolf meets the Mississippi in downtown Memphis) and to the event organizers and volunteers—y'all did an amazing job! Y'all are River Angels in my eyes.

And last, but certainly not least, good ol' Dale Sanders one-upped his rival, John Henry, by becoming the first person to canoe the Wolf River end to end in one continuous journey. As of this writing (April 2016), his accomplishment has not been repeated, nor attempted. Not too shabby for a grey-bearded 76-year-old.

Dates, distances, and other details from the SUPAS expedition (stats courtesy of Mike Watson):

DATE	START TIME	ELAPSED TIME	MILES	SPEED	MILE MARKER (CAMP)
4.7.12	9:01:35 AM	9:48:36	5.2 mi	0.5 mph	96.4
4.8.12	10:14:24 AM	7:38:21	6.2 mi	0.8 mph	90.8
4.9.12	11:13:08 AM	8:45:29	9.0 mi	1.0 mph	82.9
4.10.12	8:56:30 AM	9:16:23	11.7 mi	1.3 mph	71.5
4.11.12	8:07:24 AM	10:01:52	18.1 mi	2.0 mph	55.0
4.12.12	10:00:30 AM	8:22:03	16.5 mi	2.0 mph	38.8
4.13.12	8:10:04 AM	9:52:10	22.2 mi	2.0 mph	17.3
4.14.12	7:57:14 AM	7:38:25	18.6 mi	2.0 mph	—

Cell phone reception had been meager at best on the upper Wolf, so keeping up with the Bikecar shipping details was next to impossible. We had no idea whether the bike was on its way to Memphis or if it was still sitting in Eugene, Oregon. As we neared civilization on Day 5, I was able to retrieve an email from Paul Adkins in Eugene.

RIVER ANGELS

On April 11, 2012 at 8:21PM, Paul Adkins wrote:

Rod,

All systems go. The Bike Car has been picked up.

Paul

 A voicemail from BKK Transport confirmed that the bike was on a truck and heading east. With any luck, it would be waiting in Dale's garage when we finished SUPAS on April 14.

 As hard as I had tried to keep the cost of shipping to a minimum, the early offers of cheap cartage never bore any fruit. Shipping companies never followed through on their bid offers. Drivers never showed up when planned. Phone messages went unreturned. The whole ordeal was incredibly frustrating. By the time expediency became an issue, the shipping costs had tripled and were now well beyond what Dave Cornthwaite could afford. The Bikecar expedition was in jeopardy of being cancelled.

 To help offset the cost, I offered to pay half of the $1500 fee. In exchange for my "donation" to his expedition, Dave offered to help design the cover for my first book and possibly do some film editing for me. With Dave wrapped up in the Honolulu airport fiasco, I fronted the money for the shipping with the promise that Dave would reimburse me later.

 April 15th and 16th—the two days following the SUPAS expedition—were busy affairs. First off, the Bikecar was not sitting in Dale's garage when SUPAS finished. Calls to BKK Transport ensued and we discovered that the bike was parked on an incapacitated truck somewhere in Salt Lake City, Utah. The company offered apologies and excuses, but no concrete arrival date in Memphis. We would simply have to sit and wait.

 Corn spent most of the 15th editing a 12-minute documentary about the SUPAS expedition. The 16th saw Dave doing four school talks in the Memphis area. I tagged along for the day, sharing stories about snakes, spiders, and the hardships we had encountered in the swamps of northern Mississippi. Kudos to Ken Kimble, Outreach Co-ordinator at the Wolf River Conservancy, for shuttling us around that day.

The school talks had been organized by the Wolf River Conservancy as a way of raising awareness about the river to local schoolchildren. The WRC had been instrumental in getting Dave to Memphis from Hawaii by offering to pay him for the speaking dates. Although not confirmed, I believe they also helped pay for his flight. In any case, there was a lot of mutual backscratching going on. Corn certainly used his "celebrity adventurer" status to the max in order to benefit himself. He was good at it. He'd had plenty of practice.

I admit to being jealous of how much attention Dave received versus the amount that came my way. I had played second fiddle to him on the Wolf (much to my dismay) and felt the same kind of resentment while watching how things played out post-SUPAS. Corn seemed to have a giant horseshoe jammed far up his ass. His luck was amazing. Or maybe, it was just hard work paying off. In any case, I felt that the amount of attention paid to Corn, and the amount of rewards bestowed upon him, were disproportionate to what I was receiving. I expected more and was disappointed when it didn't arrive. Courting the illusionary beast known as *entitlement* is indeed a dangerous business. Satisfaction derived from such a relationship is fleeting at best. A far more common outcome is frustration. *Heaps* of frustration.

Dave's eagerness to disembark on his Bikecar journey was never evident throughout SUPAS or the post-SUPAS events. If it was, I missed the cues. He made no mention to me about his hidden agenda of paddling with his friends in the 22-mile Bluz Cruz race in Vicksburg, Mississippi. He had hoped to arrive in Vicksburg in time for the race on April 28, but now, with the Bikecar shipping delay, his race plans were in jeopardy. It wasn't until he posted a short expedition update video to Facebook that I clued in to his silent frustration of having too much downtime. Here's a quote from the video:

"This is the driveway. I've been staring down it for days. Waiting. Waiting. Waiting for a Bikecar. I should be riding now. Memphis to

Miami. 1001 miles on a Bikecar. It should've been here a week ago. Truck broke down. I'm bored of waiting."

Overall, the video had a humorous tone—typical of Corn's videos—but it was also permeated with an unspoken unease. Corn's underlying agitation triggered a feeling of guilt in me, due to the fact that the shipping delay had been partially my fault. I had promised to undertake the task of getting the Bikecar from Eugene to Memphis. I'd succeeded, but the achievement had cost us time and money. To reconcile, and to help alleviate my guilt, I had offered to pay half of the shipping. Doing so rooted me deeper in a situation that was making me feel more awkward and insignificant by the day. For the sake of both mine and Dave's stress levels, the Bikecar could not arrive soon enough.

Adding to our stress levels was the fact that we also had to deal with the repercussions of a very head-shakingly odd thing that occurred on Day 5 of SUPAS. On the day in question, Rachel Sumner was en route to the Wolf River with a 12-foot stand-up paddleboard wedged in the rear of her speeding Jeep. While travelling down a four-lane, divided highway, the board, which was not likely tied down, took leave of the Jeep and came to rest on the roadside. For reasons known only to Rachel—perhaps personal safety being the top reason—she motored to the next exit instead of reversing on the shoulder to retrieve the board. She doubled-back in the opposite direction, spied the blue SUP across the lanes of traffic, raced to the next exit, and then sped back to the point of jettison. There was one thing wrong however—the board had disappeared. During her efforts to return to the scene of the wayward craft, someone had scooped up the SUP and made off in haste, which then, of course, left Rachel in the awkward position of having to explain to John Ruskey that one of his boards had gone AWOL, or, more correctly, had been *stolen*.

The whole ordeal was bizarre, to say the least. John Henry suggested to Rachel that she report the incident to the police immediately, and then relay the bad news to Ruskey. For reasons known only to Rachel—perhaps she feared bad press for SUPAS—she held off contacting the police and the insurance company associated with Operation Broken Silence (although she did contact them later and filed a police report).

As we found out later, she also held off contacting John Ruskey. In fact, Ruskey knew nothing about the missing board until it was brought to his attention by Dave Cornthwaite *five days* after it went missing. How is it possible, we wondered, that Rachel had somehow neglected to contact Ruskey? This didn't seem in line with her prudent character at all. Her oversight reflected badly on us and on SUPAS, but especially so for Corn. The mishap threatened to taint the friendship between Dave and Ruskey. Corn, it seemed, was not a happy paddler.

On April 17, 2012 at 1:06AM, Dave Cornthwaite wrote:

Hi all,

I emailed John Ruskey this evening to find out when he needed his boards to be delivered back to him, and also informed him about the missing board.

It seems one paddle is still missing. It was last seen going into John Henry's truck. Can somebody please liaise with John Henry on this, we can't deliver back to Ruskey without this paddle.

Also, it would be nice to let John Ruskey know how far through the police report/ insurance claim we are so he can plan for when a new board will be delivered to him. Out of respect for his generosity on this it would be nice to get him as full information as possible, especially as his remaining boards are now covered with random signatures!

Cheers

Dave

On April 17, 2012 at 6:47AM, John Ruskey wrote:

Hullo Dave! Hullo Dale! Good to hear from you guys.

Alright, alright, how was it? Did y'all have enough water to start in the headwaters? I was wondering how everything was going. Has there been any press coverage of this? I've been looking.

Dave, I'm leaving tomorrow for another 2-week run, so TODAY (Tuesday) would be the best day if you guys want to catch me at base. Otherwise it will have to be in 2 weeks. I'm cc'ing Dale and the others so you guys can decide. Either way is fine by me, but maybe it would be best if at all possible to go ahead and bring them on down so I can assess damages and losses and get with YOLO about replacements/repairs (which might take some time). Write back or give me call 662-902-7841 and let me know.

Jeez, that's hard to understand about the one board. You are the first to let me know. Was the driver drunk? I know you weren't there, but that just doesn't make sense. How can you lose a board out of a jeep and not notice? Obviously wasn't tied down. Or maybe he had them standing up in the back of the jeep like surf boards? Okay, well I'm glad it was a YOLO board — and not one of you guys — that got lost. Also glad the flying board didn't hit the next vehicle behind the jeep (one of my worst fears). I'll get with the others about replacement. I hope they have insurance, I can't afford it.

Okay Dave, hope to see you when the boards are returned. Otherwise, I will say godspeed and bon journee on the next adventure!

Yers, Driftwood

Dale Sanders replied:

We are on our way to the TV station for interviews. Plans are for us to come down this afternoon. Believe Rachel has been trying to reach you about the lost board details. Thanks so very much John for all the paddle board support you gave to OBS and the SUPAS adventure.

Oh yes, I forgot to mention that while the missing SUP drama was playing out via email correspondence, Rachel, Dave, and I were scheduled to do a live TV interview with WREG's "Live at 9" morning show. I joked with Rachel that she should mention the SUP mishap on air and issue a plea for its return. My joke fell flat. She stuck to the script and mentioned only Operation Broken Silence's mission and strategy plan. To her credit, her well-articulated speech was the interview's sole highlight. And as for the missing board—it was never seen again.

At the same time he was trying to smooth over relations between Ruskey and the SUPAS team, Corn also became embroiled in a spat with John Henry regarding an Aquapac drybag that John never received as part of Aquapac's SUPAS sponsorship. Dave, Aquapac's "Outdoor Champion 2012" (the company's ambassador), suggested that John contact the company directly. John, in his typically irreverent manner, did just that, and sent a heated email to Aquapac's owner, Tim Turnbull. Turnbull, puzzled by John's tone, forwarded the email to Corn in hopes of finding a resolution. Corn, unimpressed with the tone of John's email, replied with a curt tone of his own.

Up until this point, I had no idea there was drama between Dave and John. Apparently, this had been brewing for a couple days. I give Corn credit—he kept it well hid. What he did next, however, wasn't very smart.

Corn decided to share John Henry's email with other members of the SUPAS group. Although he never stated his motives, I feel that Dave wanted to portray John as the bad guy in all of this. By sharing John's email, he hoped to sway the others to *his* side rather than John's. Dave's move was manipulative. It reminded me of a Dale Sanders' tactic. Dale, of course, was right in the middle of this dispute, siding with Dave and fuelling the hate directed toward John. I, on the other hand, remained neutral which didn't seem to sit well with Corn and Dale. As if my feelings of awkwardness weren't already off the scale, this new scenario created a whole new level of uneasiness in me. I was stuck in the middle between feuding friends. The whole kerfuffle seemed completely unnecessary, but it *did* make for some good entertainment.

On April 17, 2012 at 5:18 PM, John Henry wrote:

Dave - I asked you about the drybag issue while you were on my front

RIVER ANGELS

steps earlier and your comment was that we should handle it individually. I did so. Perhaps that might have been the time to step up and deal with the matter yourself, as your contact @ Aquapack (Mr Turnbull) clearly told me that he only sent what YOU ordered for the team. Granted - I did change my order, but did so well before the final order went to Aquapack.

I personally spent hundreds of my valuable manhours running logistics for this major undertaking, and hundreds of MY hard-earned dollars in the process.

In my "Rude" email I merely asked for what I originally ordered, and explained what I believe is a design flaw in their SLR bag. I am not a diplomat, and always try to call em like I see em.

As for the speedy refund on the SLR bag - even that requires more logistical pain in my ass, as I never received a pro-deal code (although I emailed Aquapack several times about the issue), and was forced to ask Jonathon to order one for me. I will necessarily have to return the bag to Jonathon, and hope that he doesn't mind returning my cash...

Aquapack's sponsorship to SUPAS was GREAT and all, but, honestly - to start this expedition without the backpack which I ordered put a bad taste in my mouth.

I have some great images of not only Aquapacks gear Hard @ work in the Wolf River Headwaters Region (one of the toughest areas anywhere on earth), but also of the team holding their banner at the finish of the expedition.

I am a photographer, and before I turn over my images - would insist that I receive the one piece of gear which I actually ordered. The expedition is now over, and I still have not gotten a wet/dry backpack.

As for the relationship with sponsors - the three cases of Wine (from a company who flatly told Rachel NO when asked about sponsorship), the Two *kegs of beer (from ANOTHER company who flatly told Rachel NO when asked about sponsorship), the keyboard player ($100.00 out of my pocket) with Barbara (who would normally make at least $500.00 for her donated time behind the microphone), the sound system, the power to run that sound system (Live From Memphis), the two mayors for photo ops, the free meal @ Wolf River Cafe, the spot on "Live @ Nine", etc., etc., - that ALL came about because of MY personal efforts, and the favors which I called in on behalf of OBS. I went above and beyond the call of duty in order to ensure the success of SUPAS. I don't need accolades, or time on the air, but would expect the one piece of gear which I asked for up front.*

Everybody - (Rachel, Jonathon, Robinson, and finally - you) who I asked about the drybag issue, all acted like I was being too demanding, and should be happy with what I got. Fuck that! I ordered a Backpack and, by God - I have a backpack coming to me!

OBS's relationship with ME has been damaged to the point that I will NOT be on board for anything they do in the future. I am beyond shocked that Rachel didn't have enough respect for John Ruskey to call him herself and tell him that she lost his brand-new paddleboard! That is SHAMEFUL, and, as such - I will NEVER do anything for them again! I am sure that they have not considered that in their decision-making process - and am quite OK with that too...

To finish - I am sure the team could have done this EPIC trip without me on board, but, I am also sure that their suffering would have been much worse in the Headwaters Region. Either way - it is what it is...

Cheerio on your next expedition... :-)

Very Respectfully - John Henry

Robinson Littrell from Operation Broken Silence responded to John's email by sending a private email to everyone in the SUPAS group—*except* John. He then followed that email with a differently worded email addressed to John and everyone else. Both emails are included below.

Everyone, John has been great for the team, so please everyone be sure to thank him. We really couldn't have done it without him.

However, his actions recently and his Emails which have been unwisely worded to Aquapac have frustrated those involved, and I personally feel that it is very... uncool of him to behave like this over a mix up of orders. That being said, let's all hold our further judgements to ourselves and leave john henry alone.

He does deserve an extraordinary thanks for what he has done because he was a vital part of SUPAS.

Please everyone join me in thanking him and let's celebrate each other and all our hard work rather than harp on a messed up shipment or who got the sponsors to come out. At the end of the day, let's remember why we did what we did- whether its for the victims, or for the adventure, or for the advancement and service of God's kingdom.

I know that I could NOT have done this without all of you. Thank you so much for your dedication and hard work.

(I found Robinson's forthright mention of "the advancement and service of God's kingdom" to be very unsettling. Being privy to such displays of servitude rhetoric incites a number of feelings inside me including anger, fear, and sadness. I cannot for the life of me understand why people choose to believe in a deity that does not exist. All I can do is simply shake my head at their stupidity. America's Bible Belt is a *weird fucking place*. I really must spend less time there.)

Robinson's second email:

To john and everyone else involved in this paddle:

Thanks for all you've done, both individually and as a collective group.

To john: im sorry that your relations with OBS are irreparable. I did not realize that your feelings had become the way they are. But thank you for your scouting trips and all the networking and connections you made.

To everyone: congrats on playing a role in the first ever full length wolf river paddle! Let's not mar the incredible accomplishment that we undertook in the name of social injustice! We really made an impact that will continue to ripple outward.

Mike: ill get you back your mosquito netting as soon as we work out a way to do so

Dave: thanks for coming across the pond and participating. It was great getting to know you!

Dale: your hospitality and knowledge of the river was invaluable!

Rod: great week, eh? I had a great time paddling with you after sundown and wondering if we would make it out alive!

JB : man, thanks for the idea and fir helping us make SUPAS into a "paddle with a purpose".

Richard: thanks for helping promote and helping by getting the bccc guys and girls involved.

Luke: I didn't get a chance to say hey but I know where ill be this summer- on a sup memphis board!!

Again, thanks everyone for the opportunity to be a part of this and thanks for making SUPAS what it became.

In an attempt to set the record straight about both the missing paddleboard *and* John Henry's drybag, Rachel Sumner replied with this email:

Guys,

I'm sorry that any of this had gotten to this point. I'm extremely grateful for all of your hard work, John.

I don't believe I ever acted in any way about the aquapac gear, much less with the assumption you should be happy with a mixup. You asked me to hold the incorrect bag in a safe place to handle when the dust settled from the expedition. I still have it safe and secure. I was never a part of any of the ordering and know nothing about any of it. The mixup has been on the back burner as I've still been spending my non Hueys hours since Saturday returning items to owners/distributors, and with rape clients. I did ask JB about it last night and began to figure out what I should do to correct this even before today's emails. I'm sorry my multiple jobs may give the impression of indifference. I try to balance these responsibilities, but fall short more times than I'd like.

As for John Rusky, please, don't assume or assert to others that I didn't contact him myself. I called multiple times and different numbers. I preferred to speak directly to him rather than voicemail, and we continued to play phone tag. He's emailed me already this morning noting that he now understood why I kept calling. I have also spoken to him today, and am working to correct my mishap.

Guys, I can't thank you all enough for all of your efforts and sacrifice.

I'll be in touch with you all soon,

Rachel

As much as I wish I could print Dave's original email in which he referred to John Henry's correspondence as "extremely rude," I cannot. I wasn't on the mailing list. Maybe that's a good thing. Here, though, is John's response to Corn's original email:

Dave - in my "Extremely Rude Email"... I asked if there was some trick to making the SLR bag work. I did not ask for a refund. I personally don't feel my email to Aquapack was rude @ all. They sponsored our trip, and so did I. I have no problem acknowledging their sponsorship, but not before I receive a Wet/Dry Backpack.

I had to use their competitor's drybags - since I didn't get what I ordered from Aquapack. That is YOUR fault, since YOU placed the order. (as per Mr Turnbull).

I will be glad to provide Aquapack with some Great PR images of their gear and banner - as soon as I receive the wet/dry Backpack which I rightly ordered .

Jonathon bought the SLR bag, and as such - I will gladly return it to him.

As far as Acknowledging their sponsorship - I am wondering exactly how much their sponsorship was? I spent several hundred dollars myself on assorted gas for logistics, music at the party, etc (NOT including my personal costs). My sponsorship has not been publicly acknowledged except as my logo was put on the website. I thank Jonathan for that consideration. He is a Great guy!

I totally don't care what you think of me, as I feel that you are a manipulative player, who is used to having his way paid by others!

As far as "problems" - not sure where you were going there, but – I asked you about the drybag. You told me to deal with it "individually". I did just that, as soon as you left my house. To harsh me for doing as you said don't sit well with me... We can fight over it if you'd like! Just say when and where! :-)

Otherwise - Good luck in life! :-)

John Henry

Dave Cornthwaite replied:

The joy continues!

In an obvious effort to smooth over relations with Aquapac, Rachel Sumner sent the following email to the company's owner, Tim Turnbull.

On April 21, 2012 at 10:07AM, Rachel Sumner wrote:

Tim,

My name is Rachel Sumner, and I'm on staff with Operation Broken Silence. The event ending last week here in Memphis - Stand Up Against Slavery (SUPAS) - was to benefit the safe house we're working to open this year for minor victims of sex trafficking in this region. I'm endlessly thankful for the number of men that chose to be partnered for this event and for the victims this event would benefit. To this point, outdoor events and events with male organizers have been rare, at best, in the anti-trafficking world. SUPAS has laid a different kind of foundation for the girls, this house, and even the fact that these men have chosen to say that they would be a part of the fight.

As SUPAS was in very beginning of planning stages, you guys - AQUAPAC - were the first major sponsor on board. Your sponsorship propelled our hopes that the expedition was even possible. Without dry bags, iPad covers, even Dave's personal Aquapac gear, etc, we would not have been able to accomplish what has been accomplished.

In light of your sponsorship, and the excitement of the event, the ball took off down the mountain. We, at Operation Broken Silence, did not follow up with you the way we should have. We should have followed up with gratitude, with updates about the progression of the event, and least of all, with info about the timeline you could expect your photos. This communication would have been common courtesy, and I missed it. Now, following the event, there has been other communication that has left a bitter taste, and I apologize about that as well.

The dry bag in question was passed to me at the beginning of the expedition for safe keeping. I have received the email from Elly, and will be putting the bag in the mail this upcoming week. Honestly, your gracious reply and willingness to exchange humbles me even more of the integrity of you and the company you represent.

Thank you for being a part of the first ever, Stand Up Against Slavery here in Memphis. We could not have done this without you!

Yours,

Rachel Sumner

As for photos, there are still being compiled and edited. They are forthcoming.

John Henry eventually received his Aquapac drybag, free of charge. Thankfully, no punches were thrown between John and Dave Cornthwaite. I don't think they've talked since that incident. John also severed ties with Dale Sanders, Rachel Sumner, Operation Broken Silence, the Bluff City Canoe Club, and the Wolf River Conservancy. I guess you could say John Henry became a lone wolf. Then again, he always had been.

But even a lone wolf needs a little company now and then. It seems they can't leave things well enough alone. Call it boredom. Call it a need for attention. Call it what you will. Whatever the case, John Henry refused to let his feud with the Wolf River Conservancy fall by the wayside. He continued to taunt and critique the Conservancy's motives on Facebook, bluntly goading them over things he thought important. Eventually, they blocked him from commenting and posting on their page.

For a while, calm prevailed. John and the WRC ignored each other and life went on as it had in the pre-SUPAS days. And then, one night, something unexpected happened—something that sent an alarming message straight to the heart of the Wolf River Conservancy.

In January 2013—nine months after SUPAS—Dave Cornthwaite and I were back in Memphis for a paddlers' get-together at Dale Sanders' home. Prior to Dale's party, Corn and I met with Dale at a Mexican restaurant to share some river stories and catch up on the latest local paddling happenings. Dale wasted no time getting to the juicy bits.

"Yoo guys probably haven't heard 'bout this yet, but there was a fire at the Wolf River Conservancy office last month," said Dale. "The police think it was arson."

Corn and I turned to one another with shocked looks on our faces.

"They haven't arrested anyone yet, but they do have a handful of suspects. I probably shouldn't tell yoo this, but…" Dale paused and looked around to see if anyone was listening. He leaned in close and finished his thought. "John Henry's name came up."

Dave and I looked at each other wide-eyed.

"Wow…" I was almost speechless. I didn't want to ask the next question, but Dale's voice hinted at enough suspicion to make the question viable. "Do they think he did it?"

"Well, they're not really sayin'," said Dale. "Apparently, John was mad over the sale of a canoe that Keith Kirkland was sellin' him. It was Bill Lawrence's old canoe. Yoo might remember that John Henry talked 'bout Bill—he called him 'Yoda of the Swamp.' Bill's dead now, but apparently he knew a lot about the upper Woof. As for John—I ain't sayin' he did it, but I *do* think he made things bad for himself by sayin' all those mean things on Facebook. He brought this all on himself."

The whole scenario was stupefying. *Did John really torch the WRC?! Would he actually do such a thing?!* Based on John's Facebook comments, it certainly seemed that he was pissed at the Conservancy. *But enough to burn the place down?!* That seemed highly improbable. Even still, I had my suspicions, and I felt very guilty for thinking John was involved.

As I was rolling the mindboggling news around in my head, Rachel Sumner entered the restaurant and joined us at our table. Her presence made me feel even *more* uncomfortable. I was sitting at a table with three people who, for reasons of their own, had been entangled in some incident of drama with John Henry. They had all purged him from their social circles and wanted nothing to do with him. I, on the other hand, had no issues with the guy, other than the unnecessary hurrying and stress he'd dumped on me during SUPAS. As far as I was concerned, that shit was muddy water under the proverbial bridge. I had no other grievance with John and refused to be swayed by Dale, Dave, and Rachel's disdain for him. I wasn't going to take sides. My goal was to remain neutral. Unfortunately, doing so left me awkwardly in the middle, and I hated that it had to be that way. John hadn't been invited to Dale's get-together. Nor did I feel comfortable inviting John as my guest. Meriam, Dale's very conservative wife, was unimpressed with John's unrestrained use of expletives the last time he'd visited the Sanders' home. She made it clear to Dale that John was no longer welcome in their home. And seeing that they were offering me accommodation at their house, I felt I had little to say in the matter. I was worried that if I took John's side, or showed any allegiance to him, I'd be ostracized from the Memphis paddling community. I didn't want to lose that connection with these people. Nor did I want to lose my connection with John. I was stuck in the middle with no way out. It was a lose-lose scenario and I was *pissed* at everyone involved. It stank of childish schoolyard drama—pointless, hateful, fearful. It was something people enjoyed watching, but didn't enjoy being part of. Everyone wanted to spectate, but no one wanted to participate. It was like a fucked up reality show where the home audience cheers on their chosen hero, and *jeers* their chosen villain. And on that night, at that table in that Mexican restaurant somewhere in Bartlett, Tennessee, I shared a meal with three friends who loudly jeered their chosen villain while I quietly cheered my unsung hero. If John Henry was guilty, then we were *all* guilty. Guilty by association. Guilty of being *mean*. Guilty of being *fearful*. Guilty of being *narrow-minded*. Guilty of being *human*. The offices of the Wolf River Conservancy were burned for good reason. Their purge helped flame away the thinly disguised veil of hate that existed between the WRC and John Henry, between Dale and John Henry, between Dave and John Henry, between Rachel and John Henry, between *me* and John Henry. Our inability to empathize with the alleged villain needed razing as much as the offices of the WRC did. Someone wanted to prove a point. Someone succeeded.

On December 19, 2014, two years and a day after the Wolf River Conservancy fire, I interviewed John Henry at a Mexican restaurant in the city of Millington, a suburb just north of Memphis. We discussed the events leading up to the fire at the WRC as well as his association with Keith Kirkland, former Director of Outreach for the Wolf River Conservancy.

In late 1995, John attended a fundraising concert for the WRC in hopes of finding someone who knew more about the upper Wolf River than he did. John soon realized that few people, if any, were truly familiar with the upper river.

"I asked the lady at the door who was takin' the money. I said, 'Point out the person in this room who knows the most about the Wolf River.' She pointed at this guy, and it was Keith Kirkland. So I went ova thare an innerduced myself and I said, 'How much of the Wolf River have yoo canoed?' And we talked, and within two minutes he knew I had canoed more of the Wolf River than he had. I said, 'I did it from Blackjack Road.' He couldn't believe I had dunn it! That's the only reason I was asked to help with SUPAS—because he realized I'd canoed more of the river than him."

Besides getting him involved with SUPAS, John's long-time association with Keith Kirkland yielded yet another windfall: a friendship with the "Yoda of the Swamp," Bill Lawrence.

If anyone knew more than John Henry about the upper Wolf River, it was Bill Lawrence. In 1993, Bill self-published a paddling guidebook for the upper reaches of the Wolf, complete with hand-drawn maps. ("And thay're pretty damn accurate!" adds John.) Bill assisted Gary Bridgman and Bill FitzGerald when they descended the river—on foot and in canoes—in 1998. Keith Kirkland, joined them on that trip as well. Gary, Keith, and the two Bills had known each other for years. They shared a passion for canoeing and a love for the Wolf. Together, they brought plenty of positive public awareness to the river and helped preserve it for future generations.

Bill Lawrence died in 2011. John Henry never met him face to face, but the two did speak on the phone several times. John held Bill in high regard, often referring to him affectionately as "Yoda of the Swamp" and acknowledging his wealth of knowledge about the upper Wolf.

Shortly after his death, Bill's brother, Tom, donated Bill's favourite canoe to the Wolf River Conservancy. Knowing that John would appreciate and use the boat more than the WRC, Keith Kirkland offered the boat to John as a gift. In exchange for the canoe, John would provide assistance to SUPAS in the upper Wolf sections.

In mid-February, 2012, John was paddling with Dale Sanders on a SUPAS scouting mission when talk turned to John's recent acquisition of Bill Lawrence's canoe. John recounts his conversation with Dale:

"It was about *17 degrees!* There was *ice* on the river! The boat was sketchy as *fuck!* I was nervous. Hell, I was drinking *rum!* I'm 'bout drunk befo' we're *dunn!* And I told Dale I was buyin' the boat for $300. Well, Dale dunn run the boat through the computer and figured out it was a $1800 boat. He went to Keith Kirkland an' told 'im, 'I'll give ya 350 for the boat.'

About three weeks later, I git a email from Keith sayin', 'Thare are Wolf River guides willin' to give more than $300 for the boat. We're gonna have to git $300 for the boat.'

When I read the email, man—I saw *red!* I was like, '*Yoo muthafucker!* Yoo alreddy *gave* me the boat, and now yoo want *$300* for it! You're sayin' people'll give ya 350!'"

Although he was livid at Keith's request for payment and Dale's underhanded attempt to wheedle the boat away from him, John relented and agreed to pay for the canoe.

"I said, 'Yeah, I'll give ya $300 for the boat, but I can't do it before this fall. My money's funny. I won't have the money before this fall.'"

That email exchange, coupled with John's less than stellar association with the Wolf River Conservancy during the SUPAS expedition, soured the relationship between Keith and John Henry. John felt betrayed that Keith would consider selling the boat to Dale Sanders, even though it had been gifted to John. Heated comments surfaced on Facebook. Gossip was hot on the collective tongue of the Memphis paddling community, with Scapegoat John dodging—and dishing out—the brunt of the abuse. A dangerous fire was being stoked by all involved and things were threatening to boil over at any moment. Smartly, Keith exercised some patience. He allowed John to keep possession of the boat and waited for him to pay the money.

Months later, with still no payment made, Keith decided to contact John directly.

"He sends me a email one day sayin', 'We're gonna need the money by the end of the month.' At that point, I wanted him to *fuck off*. I was like, '*Yoo muthafucker!* Yoo *gave* me the goddamn boat, and then you want money fo' it! And now yoo want it by the *end of the month! FUCK YOO!* Go t' *court* an' git a court order, *goddamn it!* I'll pay yoo $300 this fall like I *said*. And otherwise, fuck *YOO* and *FUCK THE WOLF RIVER CONSERVANCY!!* Come *git* the boat with a court order!!' At that point, I didn't care if I was the enemy of the Wolf River Conservancy or not."

Thanks to an unseen influx of cash, John decided to go to the offices of the WRC on December 17, 2012 and pay his debt.

"Outta the blue, I had the money in my pocket one day and I thought, 'Man, I'm gonna ride by thare and pay these muthafuckers for this boat. And I did. Keith wasn't thare. I dropped the money off. 300 bucks. Gave it to tha gurl workin' the front. That night, somebody burnt the whole *fuckin'* place to the ground! *And* they stole my 300 bucks! The WRC never got the money!!"

On the afternoon of December 18, 2012, the day the Wolf River Conservancy burned, John Henry received a phone call from Keith Kirkland. After all the drama that had ensued, John was surprised to see Keith's name appear on his cell phone.

"He ain't never called me never in the world! And I see it on my phone, "KEITH KIRKLAND." And I don't know if he knows I got caller ID or what, but I got caller ID! And I answer and say, 'Hey man, what's goin' on?' He said, '*What're yoo doin'?*' And I said, 'Funny yoo ask, Keith. I'm ridin' back from Nashville in a car driven by yer good friend, Jerry Crosby! We're comin' back from a Titans game!' And Keith said, '*Really?! No shit?!*' And I said, '*Yeah, no shit!*'"

Keith talked briefly with both John and Jerry, wished them a safe trip, and then hung up. John thought the whole conversation was strange, yet meaningless. Interestingly, Keith made no mention about the fire at the Wolf River Conservancy.

Not long after his brief phone conversation with Keith, John received an illuminating text from friend and fellow paddler, Ray Graham. Suddenly, the reason for Keith's call became very clear.

"About 30 minutes later, I got a message from Ray Graham. It said, 'The Wolf River Conservancy burnt to the ground last night.' And when I saw that message, I thought, '*Goddamn!* Keith Kirkland thought I was a *suspect* and called me to see what I was doin'! He *never* said a word about the fire! He thought I burnt the Wolf River Conservancy down over that 300 bucks!"

I asked him how he felt about his friend's suspicion.

"How *could* I feel? I thought he had a better opinion of me than that."

I asked if he felt sad or angry toward Keith.

"I wasn't angry *or* sad. I was like, 'Yoo muthafucker! I didn't burn the Wolf River Conservancy down!! I mean, *c'mon!* I gave y'all tha money I owed ya!' And…I didn't really owe it to him! He *gave* me the fuckin' boat! And then later he wanted 300 bucks for it! And the only reason he wanted 300 was 'cause *Dale* offered 'im 350! People don't understand: me and Dale have a little dynamic goin' on. If Dale sees this video (John lifts the camcorder to his face and utters, 'Hey

Dale, FUCK YOU!'), he's gonna know—he offered Keith Kirkland 350 bucks for my *fuckin'* boat, *after* we did SUPAS!"

(When I questioned John on whether he told Dale the boat had been given to him [as he'd previously mentioned] or whether he planned to pay for the boat, he replied that he told Dale he planned to pay for it. His answer caused me some confusion. The timeline was off too. His stories were conflicting. When I asked him about the discrepancy, he explained that he told Dale about the boat on Day 3 of SUPAS [April 10, 2012] and that he planned to pay for it. This conflicts with the February scouting mission mentioned earlier in the interview, in which he told Dale that he got the boat for free. It wasn't until I transcribed the interview that I realized there was a discrepancy. I chose not to verify the story with Dale. In retrospect, it is somewhat irrelevant whether Dale knew John got the boat for free or whether he planned to pay for it. For continuity purposes, I tend to believe that John told Dale about the boat on April 10, 2012, and that he told Dale at that time that he planned to pay Keith Kirkland $300 for the boat. This theory was never verified with John, although the timeline of John's story suggests that my theory is correct. Again, the discrepancy is pretty much irrelevant, but it *did* sew a seed of distrust within me. Did John tell the truth? If so, why are there two *different* stories about the same conversation with Dale? It made me wonder what other discrepancies were evident, or *not* so evident, during the interview and whether what I was told was actually true.)

I asked John about whether he had been contacted by the police.

"No, but I'm sure they checked my phone records. Keith Kirkland had to tell them, 'I called the muthafucker! He was drivin' back from Nashville and been to a Titans game!'"

"That was your alibi?" I asked.

"I didn't *need* no alibi!! Why would I *have* one, *goddammit?!* If I needed one, I *wouldn'tve* had one! That night, I posted like 25 pictures of all these bars in *Nashville!* If they went to my Facebook profile they'd be like, '*Goddamn!* He was all over Bourbon Street!' Wait…not Bourbon Street, but whatever…Printer's Alley?…Printer's Road?…"

The street name was irrelevant, and John was quick to divert to the fact that the arsonists had been arrested and convicted.

"Them people got caught! It wasn't *me!* I didn't burn the muthafucker down! It was *parolees* from around the block! They were tryin' to burn the parole office and burned the *wrong place!* But they broke the window and stole the *$300!* That's the *funny* part! She (the receptionist) laid the money on the *desk!* What kinda *dumb ass woman* is gonna lay the money on the *desk?* …for some black guy can walk by who's pissed at his parole officer and say, "Hey! I'll burn this place to the *ground* and get that *$300* right there!' That's some *weird* shit!"

I agreed, adding that there seemed to be layer upon layer of drama that had somehow embedded itself into John's world.

"It's like a *dy-namic!* A *whirlwind* follows me!!"

I laughed and agreed. It was indeed a *maelstrom of shit*—probably more than what the average person endures.

When the conversation continued, John's voice calmed and he explained his good intentions.

"Hey, look. All I wanted to do was take care of my responsibilities as per the boat, because, when I got the boat, I thought it was *cool* that it was *free*. When he wanted $300, I thought, 'Okay, I'll pay yoo, muthafucker! Yoo don't even know what the boat is worth!' When the place burnt down, and he called me, I was sad—because now I can't really ever go out on the river with him and consider him my brother…because, y'know, he thought I *fuckin' burnt the place to the ground!! WHY WOULD I DO THAT?!* First of all, I'm not *stoopid!* Second of all, it ain't worth *THREE HUNNER DOLLARS for a federal felony!* I mean, *FUCKIN' ARSON IS A FEDERAL FELONY!!* Whoever did that wasn't thinkin' about their *future!* They got *busted!* Two weeks later, they burned the building *right behind there! IT WASN'T ME!!*"

I sat back in my chair and shook my head. I was dumbfounded. John's story was numbingly ridiculous, and his telling of it was even more so.

"The politics of what transpired are *mind blowing!*" I said.

"Yup," he replied, shrugging his shoulders. "All because I went and asked, 'Who's the person in this room that knows the most about the Wolf River?'"

His comment made me laugh out loud. The story had come full circle. John had traced it all back to the fateful day he'd first met Keith Kirkland.

And perhaps that's the one defining quirk in John Henry's extremely unique personality: his peculiar way of accepting responsibility for his actions and simultaneously blaming others for the consequences of his actions. Then again, maybe those traits are *not* so peculiar. Perhaps they're not confined to John alone. After all, most of us are guilty of such behaviour at least *once* in our lives. I know *I'm* guilty of it. But what seems to be most damaging is the way in which John chooses to voice his blame. His is a language of derision, contempt, hate, humour, anger, love, honesty, and political incorrectness. He seems completely uninterested in what others think of him, yet is obsessed with pleasing others. The man has a good heart. He rescues more stray dogs than anyone I know. His home is a veritable kennel! He's an amazing photographer and he's not afraid of hard work. He's been through a lot in his 50 years and he's still contributing positively to this world. He's not to be reviled, but he's not to be revered either. In fact, he would steadfastly prefer us *not* to revere him. He would tell us he is more devil than angel, more Hyde than Jekyll, and some would agree. But truthfully, he is no different than any of us. We all shine a light and we all cast a shadow. We are all imperfectly perfect and simplistically complex. We are who we are. We are human.

I chose to focus this section of the book on John because I can relate to him. I can relate to his artistic side, his introverted personality, his lack of patience with a world full of stupid people. Like him, I have self-esteem issues. Like him, I want to please others. Like him, I want to share the beauty of nature with others, but struggle with trust issues pertaining to the very people that I wish to align myself with. I pull them in. I push them away. I invite. I incite. And at the end of the day, I sit alone and justify my actions by accepting responsibility, or deflecting responsibility, or by unfairly placing blame on others.

I am far from perfect. John Henry is far from perfect. But we are close to being *real*, and being real is all anyone wants from us. May we move forward as friends and help one another become more *real*. The world is bored of masks and mean words. Let's show them the truth. Let's move forward toward Peace.

Ten days after it left Eugene, and one week after SUPAS concluded, the Bikecar arrived in Bartlett, Tennessee. Its journey from the west coast was disjointed, disrupted, derailed, and downright stress-inducing. Thankfully, it arrived in one piece.

Once we got the bike off the transport truck (it had been loaded on a 53-foot car trailer with several other vehicles) and into Dale's garage, we realized that some serious drivetrain repairs were in order. Prior to SUPAS, Dale and I had visited several of the local bike shops and requested free or discounted repair assistance, citing the charity aspect of Dave's upcoming ride from Memphis to Miami. (The expedition was raising funds for breast cancer research.) Outdoors Inc., a local sporting goods store and SUPAS sponsor, offered a free tune-up as well as discounts on bicycle accessories.

Also in need of repair was a fiberglass rooftop cargo carrier (usually seen on automobiles) that Paul Everitt had mounted on the back of the bike after he removed the two rear seats. Previously, the carrier was used on a trailer he'd built for a jaunt across Europe in 2010. Upon further inspection, we found that the carrier's hinge was damaged and the outer shell was cracked. Repairing it would take too much time. It would need to be replaced. Dave didn't have money for a new one, so a Facebook request for a used cargo carrier was posted. The post yielded instant results. It just so happened that Mike Watson, SUPAS mapmaker and team member, had a perfect match sitting idle in his ex-wife's garage. "Consider it a donation to the cause," said Mike.

The Bikecar was taken to Outdoors Inc. the next day where it underwent a thorough shakedown and tune-up. Unneeded bits were removed and discarded. The gearing system on the passenger drivetrain was scaled back to a single chainring, offering only three gears instead of the original 21. As much as Dave wanted company during his ride to Miami (mostly to aid in powering the damn thing—at

220 lbs. sans gear, the bike was a heavy beast!), he knew that most of the time he'd be pedalling it by himself, so extra gears and accessories would be irrelevant on the passenger side of the bike.

The competent crew at Outdoors Inc. (Kenny and Andrea) spent about two hours tweaking, tuning, tightening, prodding, pinching, and pulling the bike into rideable shape. When Corn and I finally got a chance to take it for a spin around the parking lot, we realized two things: 1) the bike was heavy as fuck, 2) Dave's expedition was going to be hard as fuck—as in, it was going to be *fucking difficult!*

Originally, I had planned to join Corn for the first week of his journey, but shipping setbacks shortened that plan to only one day. I was due to fly back to Ontario on April 23 to resume the planning of my kayaking descent of the Missouri-Mississippi river system, which had been put on hold in order to help with SUPAS and the Bikecar expeditions. Corn was eager to hit the road as soon as possible. He wanted to make up for lost time and was pushing hard to leave Memphis on April 22. As much as I wanted to pedal into the Deep South with Corn, I needed to get on with my own affairs. So, as things turned out, I would only join him for the first day. After that, Corn would be powering the four-wheeled beast by his lonesome, at least until he found another pedalling partner.

With the drivetrain shifting issues smoothed over and the new cargo carrier soundly secured on the bike's rear, it was time to deal with the dilapidated driver's seat. Understandably, Paul Everitt's five-month, 5000-mile trek across Canada and the northwestern U.S. had beaten his hand-crafted driver's seat into a deplorable, unrepairable state. It needed to be replaced.

Thankfully, Tom Roehm, a friend who had helped house Dave and his brother Andy during Dave's Mississippi River descent in 2011, offered to customize a new seat and refurbish the seat mounts. It was time to move beyond Paul Everitt's in-the-field duct tape and hose clamp repairs.

The Bikecar was loaded on Dale's car trailer and hauled to Big River Engineering and Manufacturing, Tom Roehm's medical devices manufacturing business in downtown Memphis. A handy ground-level loading bay provided a perfect place for Tom to work on the bike. An office chair was stripped of its pivoting pedestal and mounted on the bike's aluminum frame. As a backup—in case the office chair failed—Tom fashioned a folding lawn chair into a suitable replacement and stowed it away in the cargo carrier.

While at Big River, Dave took a few minutes to document some of his concerns about the upcoming expedition. In a video later posted on YouTube, he explained his complete ineptness where bicycles and bicycle repairs were concerned, and then turned his attention to a more telling fear.

"I'm most worried about traffic safety. Americans have this horrible habit of texting whilst driving and the Bikecar's quite wide—it's five foot, seven inches. So, anyone not paying attention could hit it, despite the fact that there's loads of visible flags fluttering above us."

The banks of the Mississippi River seemed like the best place to start the journey south, so we hauled the Bikecar down to the visitor's information center at the foot of Jefferson Avenue, not far from Memphis' famous glass pyramid and historical Cobblestone Landing. An elevated monorail towered above our heads, carrying tourists across a harbour to Mud Island. (Curiously, the harbour used to be the original course of the Wolf River. The river was diverted west in the middle of the 20th century and now empties into the Mississippi River at the north end of Mud Island, at the spot where SUPAS concluded. The diversion created a harbour which is now filled with the fluctuating flow of the Mississippi rather than the languid drift of the Wolf.)

After the obligatory round of send-off videos and photos (courteously snapped by Rachel Sumner), we set the Bikecar wheels in motion. SUPAS teammate Richard Sojourner joined us on bicycle (as did Tom Roehm) for the first mile as we struggled up a short hill and slowly made our way south through the busy Memphis core, all the while attracting curious stares from pedestrians and drivers alike.

Our route down Front Street soon switched to Third Street and we paused briefly to bid adieu to Tom and Richard. Dale followed close behind in his van, the hazard lights on his trailer flashing repeatedly. Dale promised to stay with us until we were clear of the downtown traffic.

Outside downtown, we gained use of the curb lane and passed by the gritty neighbourhoods and run-down commercial buildings of South Memphis. In the heart of this crime-ridden community, our progress was halted when the driver's side drivetrain became a wobbly, rattling mess. Somehow, four short bolts had worked their way loose on the chainring assembly and now lay lost on the pavement somewhere between us and downtown Memphis. With no spare bolts in our repair kit, I chose to remove three bolts from the passenger side chainring—leaving the passenger side with only two bolts in the five-bolt pattern—and use them for the driver's side. The swap worked, but not before the curious owner of a greasy garage came out to see what we were doing.

"Everythin' okay here?" he asked.

"Yup, fine thanks," said Dave. "Just making some repairs to our Bikecar."

"Yer on private property, y'know," said the man, seemingly unimpressed with Dave's British accent and his funny looking four-wheeled bicycle. "Yer gonna hafta move along. You cain't stay here."

"We'll be on our way in just a few minutes, sir," said Corn, trying to muster up as much politeness as he could. "If you can give us a minute, we'll be out of your hair."

The man gave both Dale and me a silent, once-over glance, sneered, and returned to his shop. I hurried the repairs along and moments later were rolling again.

During our repair break, Dale offered to stay on as support driver for the remainder of the day. We gladly accepted. In light of what was about to transpire a few miles down the road, this ended up being one of the best decisions of the whole day.

As we neared the state line and crossed into Mississippi, Memphis' urban sprawl gave way to horizon-wide views of cash-crops and four lanes of arrow-straight highway. We powered along at a steady 10mph as a warm headwind breezed out of the south. Talk was minimal. Our goal was to cover as much ground as possible before I departed.

The rhythm of the road and the repetitive movements of self-propelled travel lulled us into a familiar state of meditation. Our minds were free to wander from subject to subject, from past to future and back again. We could plan or reflect or observe the wide-open world around us. We could make amends with old lovers, think about ones to come, listen to birdsong, or compose a symphony. As long as the pedals spun and the tires turned clockwise, as long as our muscles flexed in actions familiar and our awareness remained half-trained on the task at hand, all was well. Expedition #6 on Dave's list of 25 was progressing. He was happy. I was content. Dale had our backs. All was good. And then, everything changed.

The horrible, screeching sound of angry resistance hit my ears hard without warning, followed a millisecond later by a sickening metallic crunch that foretold only one thing: impact between two vehicles. The noise jolted me from my trance and my head spun left in search of the source. Car tires tore into pavement. White smoke rose from hot rubber. The screeching increased to a feverish pitch. It filled my ears and purged all pure thoughts from my meditation. Attention became acute. The present moment stretched, slowed, and waited for me to acknowledge it. I did. It came at me with a vicious sneer—angry and unforgiving. If I'd done anything wrong in life, if I'd hurt others with toxic words and disparaging actions, surely this was the inevitable payback, this was the inescapable punishment. Guilt washed over me like a rogue wave—unexpected and unwelcomed. I stared long at fate disguised as a silver sedan skidding sideways towards me a 60mph. It was unstoppable. In short moments it would compress the warm air between us and push that air through my gaping mouth and into my empty lungs. Organs would convulse. Cells would hemorrhage. A life of movement would cease. A new trajectory would be instated, one whose destination represented radicalized transformation and consummate decomposition.

Forward it came, like rabid rats charging through a cat door—teeth bared, eyes wild, skin-cringing cries bursting forth from dry throats, fur standing on-end, tails thrashing the air, coming, coming, claws fast upon the linoleum, god-awful scratching, screeching, shrieking, driven mad, they are upon me, incisors driven to the bone, tearing flesh and searing pain, repulsive like the smell of burning fingernails.

RIVER ANGELS

'Twas a petrol propelled purge. A hurling dervish. A devil delightful. It arrived right on time and time teetered precariously between sanity and the morrow. A marrow. A barrow. A wheeled-away truth. A reason. A treason. A company uncouth. A bitumen bed partner who satisfies none. A wanker. A winner. A party of one. Spread not your legs and invite those you ask, but close fast the casket and have the last laugh. Here comes a problem, a riddle, a test. Cast free your intentions, for your intentions are not best, suited for a future of 9 to 5 hatred and humourless toil. Be not the fool, the failure, the fall-guy, the foil. Be instead the direction, the action, the pith. Become seed, fruit, and harvest. Become your own myth. For here arrives a catalyst whose wicked resolve will not cease. Collision, collusion, commission, delusion. Lover, oh lover, why smother me now? A pillow we shared. A future we dared. A fortune we fared. A burden we bear.

Can I pay you to arrest these thoughts in my head? Can you attest to my pain? I want to see a candle in your navel and a halo around your head. See my smile? Know that it is *false*. Have you seen me regally preen for your entertainment? Know that justice serves only those that believe in it. I believe only in leper cons and uniforms. I believe only in the fantasy that is *you*.

Leave me as does a dream before awakening. I seek not to retain such thoughts, to entertain such pain. I seek only solitude, the comfort of my own cancer. Your presence is an unbearable pressure upon my soul. Thoughts of you drip like hot wax upon my addled brain and the ensuing, singeing stench fills my senses with a gust of disgust and a dose of dissension. Trudging through a mutinous stew, I nervously pace the pirate's plank like a man who refuses to meet his destiny, like a man who chooses to forestall the inevitable.

"You there!" cries the Captain, "Reveal your pirate parts!"

And then it hit me, hit us, hit *everything*. Everything became nothing and nothing came to rest in a place unfamiliar. I averted my eyes in time to miss the impact. I saw Corn staring forward, his knuckles white upon the handlebars. A sickening thud shot us sideways, out of our lane and down a sloped shoulder thick with thigh-high grass.

"Don't brake!" I shouted. "Don't brake!"

The Bikecar bumped over clumps of earth and came to rest in a field of young wheat. Behind us, back up on the highway, a car horn blared incessantly, a droning soundtrack to a sorrowful scene.

Corn turned to me, and I to him.

"You okay?" he asked.

I nodded and immediately vacated the passenger seat. Corn grabbed his GoPro camera and followed me up the slope.

The whining horn drew me in and I found myself standing bent at the waist attending to a middle-aged woman in the driver's seat of a silver sedan. A deflated airbag hung limp from the steering wheel. She appeared unhurt, but dazed nonetheless. I resolved to stay with her, to comfort her, to assure her we were alive.

Extending away from her car were two black streaks of rubber angled long across both lanes. The impact with the driver's door of Dale's van had spun her sedan clockwise 180 degrees. Skidding out of control, the car bore down on us quickly and struck its driver's side against Dave's half of the Bikecar. His left shoulder and leg absorbed some of the shock. The bike's robust aluminum frame and rear wheel absorbed the rest. Despite the frightening jolt, neither of us were seriously hurt.

Dale was still seated in his van, vigorously trying to close his driver's side door. The sedan's impact had punched a deep dent in the door, damaging it severely. Thankfully, Dale was unhurt.

Two police cruisers soon arrived on the scene, the officers taking photos, statements, and measurements. Insurance companies were alerted and litigation concerns were quietly aired. Blame was aimed. Responsibility was claimed. Causes were guessed. Damage was assessed. A tow truck hauled the sedan northbound, back to Memphis.

At a point unknown, shock befell me—a shock associated with accidents and allegations, litanies and litigations, epiphanies and exonerations. A shock that blunts the shiny day and makes it murky and meek. A shock that bloodily guts a goal and stamps out ambition. Mine was a shock that forcefully tore down a thick wall of resilience and replaced it with a withering veil of uncertainty. My shock made sure that progress stalled completely. I stood one-legged on a precipice with a strong breeze at my back and a gaping vortex before me.

I took leave of the scene and lay a head heavy with fright on a fold-down bed in the rear of Dale's van. My brain was exhausted. My body spent. Beside me sat a stuffed bear clothed in a shirt and shorts that belonged to a dead boy. Its expression was neutral, comforting, disconcerting. Jesse Nealey died at age 14 from injuries sustained in a bicycle accident. He had been struck by a vehicle driven by a family friend. It happened close to home, at a moment no one expected, at a time when teen life played in the streets and the streets lay wide with possibility. Traffic calmed. The day's dusk dawned. Pavement invited. A rider delighted with the company of friends and new love interests. An intervention arrived unannounced and broke wide a gulf between living and lost, a hard fathomed cost, a life source unplugged, a deep hole to be dug, a city to mourn, a young family torn, between fault and forgiveness, between guilt and acceptance, between grievance and justice.

A blunt reminder had been wrought by events that sought to teach the untaught a lesson of love and a glance at life's light, its promise white bright and exceptionally lucid, like how a tear makes clear a blurred out briskness, an icy avoidance, or an annulment of patience. There comes a time in life when truth arrives bloodied and bold, and its message embeds itself in our collective jugular, spilling awareness like an uncorked bottle of aged redemption. We see a world stained beyond what we knew, beyond faith, beyond conviction, beyond creed, beyond diction of deities we never have viewed, beyond dreams dreamt by dreamers who never have hued a future uncertain, a future imbued, with pride, risk, and purpose, a future *accrued*.

With richness in hand, we walk a new land, a land sewn with goodness, with helpers, with aid, with seekers and strivers and searchers who paid, for their freedom with blood, with tears, and with pain, and strove ever onward to stay half-way sane. And there at our shoulder, and there at our feet, and there on the hilltop or out in the street, on bridges, on oceans, on billboards, on trains, are angels who watch us and measure our gains. They see us on Facebook, they see us leave home, they follow our journeys, wherever we roam. They take a great interest in the things that we do, because they are, like me, not much different than you. And perhaps that is why, when they drive slowly by, and clap, wave, and smile, every inch, foot, and mile becomes easier to cross on our way to our goal, on our way to becoming happy, healthy, and whole.

They go by names like Jamie Zelazny, who just happened to see Dave's Bikecar posts on Facebook that very morning. Jamie was finishing up work at a casino in nearby Tunica and was motoring north to his home in Memphis. He kept his eyes peeled for two gents on a big bike heading south, but saw only the flashing lights of emergency vehicles. He slowed and recognized Dale, Dave, and me and reversed his direction at the next highway exit.

We met Jamie the previous summer during a talk Dave gave at a museum on Mud Island in Memphis—right across the harbour from where we began our Bikecar journey. Jamie is no stranger to adventure. He provided hands-on logistical help for long-distance swimmer, Martin Strel, on both his Amazon and Yangtze swims. An avid motorcyclist, Jamie has travelled tens of thousands of miles on two wheels. He was straddling his steed when he arrived at the accident scene near Walls, Mississippi.

Thanks to a battered rear wheel, the Bikecar was temporarily out of commission. We loaded it on Dale's trailer and pondered our next move. I was out of commission as well. Shock had set in and I found myself sobbing with thoughts of 14-year-old Jesse Nealey and the unfortunate collision that eventually claimed his young life. We had been spared. "Jesse was watching over you that day," said Billy Nealey, Jesse's father, in a later phone conversation.

In my semi-coherent state, I assumed Dave was going to pack it in for the day, return to Dale's house, repair the rear wheel, return to the same spot in the morning, and resume the journey south. It never occurred to me that perhaps he was thinking about packing the *whole expedition* in—thinking about how much safer it would be to count his blessings and move on in a different direction, a direction that didn't involve speed-happy drivers and over-wide bicycles.

RIVER ANGELS

Thankfully, he chose to absolve his fears and press on. And thankfully, Jamie just happened to have the tools on his motorcycle to fix the ailing Bikecar.

The four of us drove a few miles down the highway to the town of Tunica where we enjoyed some cold drinks and a meal at a roadside diner. We said our goodbyes to Corn, not knowing when we'd see him again, but hoping we would. In the dark blue Mississippian twilight, Jamie led Corn and the Bikecar to a quiet church on the far side of Tunica. Dave thanked Jamie for his help, and sleepily set up his tent. Day 1 was a mental and physical workout. Corn was exhausted. Little did he know what the coming weeks had in store for him. He'd passed the first test, but a dozen more lay in wait. (You can read more about Dave's Bikecar journey in his book, *Life in the Slow Lane*. I highly recommend checking it out.)

Sleep came easy for me that night. Dale's basement was a quiet shell, free of the madness that had occupied it for the past three weeks. Gone was the manic energy that permeates the fragile transition from *planner* to *doer*—those awkward moments when the anticipation of what lay ahead pulls us forward and away from the sensible anchor of comfort and planning. I had gone from planner to doer and was now back at the former. I had a river system to paddle, the longest in North America. The Missouri-Mississippi was calling me. Montana and the towering mountains of the Continental Divide were calling me. It was time to set a new course, time to search out a new source, time to follow a force familiar to me, time to follow a river, all the way to the sea. It was time to find more River Angels.

ACKNOWLEDGEMENTS

Big thanks to those who helped immensely throughout the evolution of this book:

My family for their love, patience, and understanding (my father Robert Wellington, my sister Carrie Formosa and her family, my sister Sharon McLean and her family).

Dave Cornthwaite (www.davecornthwaite.com) for being awesomely inspiring; the Memphis Crew for their love of rivers and people (Isaiah Allen, Jonathan Brown, Richard Day, Mary Finley, Ray Graham, John Henry, Anna Hogan, Genny Kilpatrick, Keith Kirkland, Robinson Littrell, Tom Roehm, Dale Sanders, Luke Short, Richard Sojourner, Sandy Stacks, Rachel Sumner, Mike Watson, September Young, and Jamie Zelazny); thanks to Ray Graham (www.wolfriverguide.blogspot.com) for granting permission to reprint his writings; thanks to John Henry (www.swamptours.blogspot.com) for granting permission to reprint his writings; thanks to Gary Bridgman, Tim Phillips at the Oxford Eagle/Oxford Town magazine (www.oxfordeagle.com), and Keith Cole at the Wolf River Conservancy (www.wolfriver.org) for granting permission to reprint Gary's writings; thanks to John Ruskey and Quapaw Canoe Company (www.island63.com) for loaning their SUPs to SUPAS and for granting permission to reprint John's writings; Maryellen Self for her boundless energy and awesome hugs; Norm Miller and Kris Walker for sharing their thoughts on paddling and life; Jessica Andrews for the name, "The Grey Truck"; Mia McPherson and Ron Dudley for pointing me toward Bill West; Bill West at Red Rock Lakes National Wildlife Refuge for pointing me toward Brower's Spring; Michele Glasnovic and her partner, Glen, for welcoming me into their home; Paul Everitt (www.goingsoloadventure.blogspot.com) for building the amazing Bikecar; the staff at Outdoors Inc. (www.outdoorsinc.com) in Memphis for repairing the Bikecar; Paul Adkins in Eugene, Oregon for storing the Bikecar; the staff at Operation Broken Silence (www.operationbrokensilence.org) for their tireless work; Barbara Blue (www.barbarablue.com) for providing the soundtrack.

Extra special thanks to all the River Angels out there.

Thanks to Candice Cottingham at Abstract Marketing for her graphic design help (www.abstractmarketing.ca).

Thanks to Victoria Colotta at VMC Art & Design for designing the cover spread, laying out the interior of the book, and assembling the e-book (www.vmc-artdesign.com).

Fret not. If you weren't thanked in this book, you'll be thanked in the next one. ☺

REFERENCES

1. From Source to Mouth…. (n.d.). Retrieved March 1, 2016 from www.flickr.com/photos/wolfriver/sets/72157594572079868.

2. New Territory: HWY 72 to Michigan City. (2012). Retrieved March 1, 2016 from Paddling the Wolf River: www.wolfriverguide.blogspot.com/2012_02_01_archive.html.

3. Wild Miles on the Middle & Lower Mississippi River (n.d.). Retrieved March 1, 2016 from Wild Miles on the Middle & Lower Mississippi River: www.wildmiles.org.

4. Baker's Pond. (2012). Retrieved March 1, 2016 from Paddling the Wolf River: www.wolfriverguide.blogspot.com/2012_02_01_archive.html.

5. The Source: Goose Nest Creek to Black Jack Road (3 Bridges). (2012). Retrieved March 1, 2016 from Paddling the Wolf River: www.wolfriverguide.blogspot.com/2012_03_01_archive.html.

6. The Last and Longest Section of the Upper Wolf. (2012). Retrieved March 1, 2016 from Paddling the Wolf River: www.wolfriverguide.blogspot.com/2012_03_01_archive.html.

7. The Final Chapter of the Upper Wolf: Pipeline #5 to HWY 72. (2012). Retrieved March 1, 2016 from Paddling the Wolf River: www.wolfriverguide.blogspot.com/2012_03_01_archive.html.

8. The Wolf, The WHOLE Wolf, And Nothing But The Wolf - So Help Me GOD! (2012). Retrieved March 1, 2016 from Where the Wild Things Are: www.swamptours.blogspot.com/2012/03/wolf-whole-wolf-and-nothing-but-wolf-so.html?m=1.

ABOUT THE AUTHOR

On April 2, 2013, Rod Wellington became the first North American to kayak the Missouri-Mississippi river system from source to sea, a distance of 3800 miles. The solo journey took 256 days to complete. Rod's triumphant arrival at the Gulf of Mexico marked the completion of Stage One of his Magnificent Seven Expedition, a personal quest to descend the longest river system on each continent from source to sea using only human-powered transportation (kayaking, rafting, and walking). He estimates it will take 15 years to complete the project. Rod plans to write a book about each of these seven journeys.

No stranger to adventure, Rod has bicycled more than 15,500 miles, including continental crossings of North America and Australia. He has also logged over 8070 miles of river travel, including source to sea descents of the Mississippi River (2300 miles) and the Murray River, Australia's longest waterway (1550 miles).

Rod is an accomplished public speaker and author. *River Angels* is his second book.

Find out more at www.rodwellington.com.

Made in the USA
San Bernardino, CA
10 August 2016